Integrated Circuit Authentication

Mohammad Tehranipoor • Hassan Salmani
Xuehui Zhang

Integrated Circuit Authentication

Hardware Trojans and Counterfeit Detection

Springer

Mohammad Tehranipoor
ECE Department
University of Connecticut
Storrs, CT, USA

Hassan Salmani
ECE Department
University of Connecticut
Storrs, CT, USA

Xuehui Zhang
ECE Department
University of Connecticut
Storrs, CT, USA

ISBN 978-3-319-00815-8 ISBN 978-3-319-00816-5 (eBook)
DOI 10.1007/978-3-319-00816-5
Springer Cham Heidelberg New York Dordrecht London

Library of Congress Control Number: 2013946790

Printed on acid-free paper

Springer is part of Springer Science+Business Media (www.springer.com)

Dedicated to my students
MT
Dedicated to my family
HS
Dedicated to my family
XZ

Preface

Outsourcing the design and fabrication of integrated circuits (ICs) has raised major concerns about their security and reliability. Realized by the intentional modification of design characteristics, hardware Trojans can obstruct a system's availability or intercept its confidentiality. Counterfeiting is also a growing issue that has raised serious concerns for the government and for industry. Counterfeit electronic components are unauthorized products that do not conform to their original design specifications. In addition to diminishing system dependability, counterfeiting reduces companies' total revenue from their research and development, discourages innovation through the theft of intellectual properties (IPs), and produces low-quality products under established brand names.

This book is intended to address these issues by presenting comprehensive and practice-oriented solutions for IC authentication. It provides insight into the IC supply chain and studies its vulnerabilities to hardware Trojans and counterfeiting.

This book is organized into 11 chapters. The first chapter provides an introduction to VLSI system integration and discusses hardware Trojans and counterfeiting as its two most challenging security issues. Chapter 2 presents a case study on the use of formal verification and code coverage analysis to detect Trojans inserted into third-party IP cores. Chapter 3 demonstrates a side-channel signal analysis technique for detecting Trojans in integrated circuits fabricated by untrusted foundries. Chapters 4 and 5 describe two design for hardware trust techniques to improve hardware Trojan detection. These techniques help activate a Trojan more effectively and increase side-channel signals induced by Trojans, making detection much easier for test engineers. Chapter 6 presents another design for hardware trust technique, an on-chip structure based on a ring oscillator network (RON) designed to monitor voltage fluctuations induced by Trojans and to differentiate noise created by hardware Trojans from noise made by process variations. Chapter 7 provides a comprehensive vulnerability analysis to effectively quantify the difficulties of the activation and observation of each circuit part. Further, it defines the Trojan detectability metric as a measurement of Trojan impact on circuit characteristics. Chapter 8 introduces the built-in self-authentication (BISA) technique as a means to prevent Trojan insertion during GDSII development and mask generation by an untrusted foundry.

Chapter 9 presents a detailed taxonomy of counterfeit electronic components, defects, and anomalies associated with each of the counterfeit types and identifies the available detection techniques. It also discusses the major challenges for detecting and preventing counterfeit electronics. Chapter 10 demonstrates design of lightweight sensors that prevent counterfeiters from recycling parts found in used electronic systems. When such parts are recycled, test engineers can detect them extremely easily with these sensors, which provide information about the chip's usage in the field. Finally, Chap. 11 discusses a fingerprinting technique for the detection of counterfeit parts using circuit timing analysis.

Storrs, CT, USA

Mohammad Tehranipoor
Hassan Salmani
Xuehui Zhang

Acknowledgements

The authors would like to acknowledge National Science Foundation (grants 1059390 and 0844995) for supporting projects in the domain of circuit vulnerability analysis, development of trust benchmarks, and hardware Trojan detection and prevention. We would like to thank Kan Xiao and Ujjwal Guin at The Center for Hardware Assurance, Security, and Engineering (CHASE) at The University of Connecticut (UCONN) for their contribution to Chaps. 8 and 9, respectively. We would also like to thank Dr. Ivy Linton Stabell who provided great assistance with the proofreading of the book.

Contents

Acronyms

3PIP	Third-Party IP
AC	Alternating Current
AES	Advanced Encryption Standard
AF	Antifuse
ASIC	Application-Specific Integrated Circuit
ATE	Automatic Test Equipment
ATPG	Automatic Test Pattern Generation
BISA	Built-in Self-Authentication
CAD	Computer-Aided Design
CAF	Clock AF
CAM	Content Addressable Memory
CMOS	Complementary Metal Oxide Semiconductor
COTS	Commercial Off-The-Shelf
CUA	Circuit Under Authentication
CUT	Circuit Under Test
DC	Direct Current
DEF	Design Exchange Format
DES	Data Encryption Standard
DFHT	Design for Hardware Trust
DFT	Design for Testability
DoD	Department of Defense
DoS	Denial of Service
dSFF	dummy Scan Flip-Flop
DUA	Design Under Authentication
FF	Flip-Flop
FPGA	Field Programmable Gate Array
FSM	Finite-State Machine
FTIR	Fourier Transform Infrared
GD	Geometric Distribution
GDS	Graphic Database System
HCI	Hot Carrier Injection

HVT	High Threshold Voltage
IC	Integrate-Circuit
IUA	IC Under Authentication
LBIST	Logic Built-In Self-Test
LFSR	Linear Feedback Shift Register
MC	Monte Carlo
MCS	Monte Carlo Simulation
MISR	Multiple Input Signature Register
MSD	Moisture-Sensitive Device
NBTI	Negative Bias Temperature Instability
OCM	Original Component Manufacturers
OCM	Original Equipment Manufacturers
ORA	Output Response Analyzer
OTP	One Time Programmable
PCA	Principal Component Analysis
PCB	Printed Circuit Board
POC	Payload Output Change
PUF	Physical Unclonable Functions
R&D	Research and Development
RAM	Random Access Memory
RCR	Redundant Circuit Removal
RO	Ring Oscillator
RON	Ring Oscillator Network
RTL	Register Transfer Level
RTOS	Real-Time Operating System
SAF	Signal transition AF
SAM	Scanning Acoustic Microscopy
SEM	Scanning Electron Microscopy
SI	Scan Input
SOA	Simple Outlier Analysis
SVT	Standard Threshold Voltage
TCA	Trojan-to-Circuit Activity
TCC	Trojan-to-Circuit Charge consumption
TCL	Tool Command Language
TCP	Trojan-to-Circuit Power consumption
TE	Test Enable
TE	Test-Enable
TIC	Trojan-induced capacitance
TPC	Test Per Clock
UNSP	Unused Spaces
VHDL	Very high-speed integrated circuit Hardware Description Language
VLSI	Very Large-Scale Integration

Chapter 1
Introduction

Human society relies heavily on computer systems. They enhance our quality of life by delivering performance, bringing accuracy, and providing security. Applications ranging from nuclear plant controls and jet engines to home appliances like dishwashers and microwaves benefit from computer systems. The dependability of a computer system determines its accountability. The dependability of a system is based on the compliance of delivered services by the system with its functional specifications. The function of the system is described by functional specifications in terms of functionality and performance. The service delivered by the system, on the other hand, is its behavior as it is perceived by its user(s). A broad concept, dependability encompasses availability, reliability, safety, integrity, and maintainability attributes as described in Table 1.1 [2].

Security is more specific, focusing on availability, integrity, and confidentiality. System security demands availability for only authorized actions, integrity with improper meaning unauthorized, and confidentiality. Trust is the dependency of a system (system A) to another system (system B), through which the dependability of system A is affected by the dependability of system B. Trustworthiness in a system is the assurance that the system will perform as expected [2].

A modern society utterly depends on integrated circuits (ICs), or chips, which are the virtual brains for all electronics. In the interest of economic matters, most companies nowadays mostly outsource and fabricate ICs overseas, rendering them increasingly vulnerable to malicious activities such as design modifications created to sabotaging a mission or counterfeiting integrated circuits.

1.1 Hardware Security and Trust

A computer system development, as shown in Fig. 1.1, consists of several steps which are not necessarily performed all in the same design house. The first step is to determine system specifications based on the customer's needs. A complex

M. Tehranipoor et al., *Integrated Circuit Authentication: Hardware Trojans and Counterfeit Detection*, DOI 10.1007/978-3-319-00816-5_1,

Table 1.1 Dependability attributes

Attribute	Definition
Availability	Readiness for correct service
Reliability	Continuity of correct service
Safety	Absence of catastrophic consequences on the users and the environment
Integrity	Absence of improper system alteration
Maintainability	Ability to undergo modification and repairs
Confidentiality	The absence of unauthorized disclosure of information

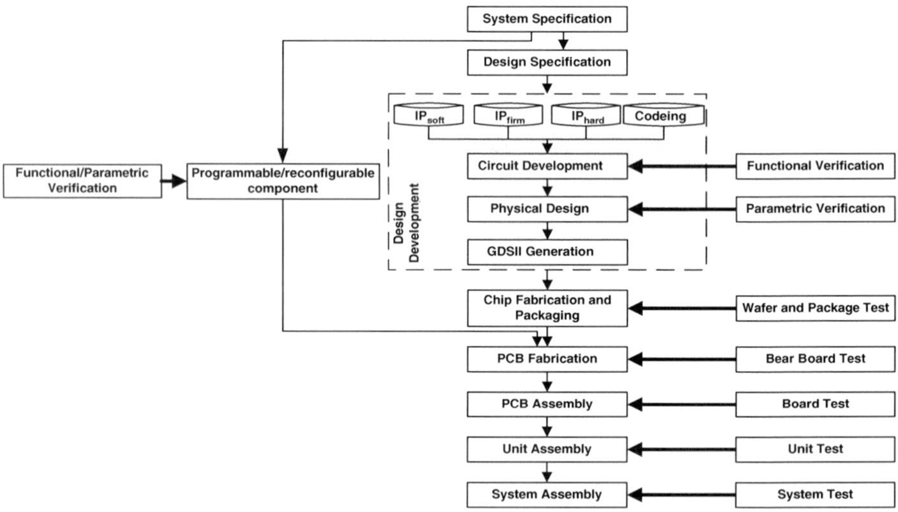

Fig. 1.1 System integration and test process

system may require a variety of components like memories and chips with different applications and functionalities.

After providing the system specifications and choosing the structure of system and its required components, design development requires different tools. Each component demands specific attention to meet all the system specifications. To expedite system development and to reduce the final cost, outsourced alternatives have gradually replaced in-house processes. Third-party intellectual property (IP) cores have displaced the in-house libraries of logic cells for synthesis. Commercial software has supplanted homegrown Computer Aided Design (CAD) tool software. In the next step, designed chips are signed-off for fabrication. Nowadays, most companies are fabless, outsourcing mask production and fabrication. Beside custom designs, companies can reduce total cost and accelerate system development by using commercial-off-the-shelfs (COTSs), reprogrammable modules, like micro-controllers, reconfigurable components, or field programmable gate arrays (FPGAs). Afterwards, they manufacture printed circuit boards (PCBs) and assemble system components on them. Finally, the PCBs are put together to develop units; the entire system is the integration of these units.

In each step, different verifications or tests are performed to ensure its correctness, as shown in Fig. 1.1. Functional and parametric verifications ascertain the correctness of design implementation in terms of service and associated requirements, like power and performance. Wafer and package tests after the fabrication of custom designs separate defective parts and guarantee delivered chips. The PCB fabrication is a photolithographic process and susceptible to defects; therefore, a PCB should be tested before placing devices on it. After the PCB assembly, the PCB is again tested to verify that the components are properly mounted and have not been damaged during the PCB assembly process. The tested PCBs create units and finally the system, which is also tested before shipping for field operation [12].

Each step of system development is susceptible to security breaches. An adversary may change system specifications to make a system vulnerable to malicious activities or susceptible to functional failures. As external resources, like third party IPs and COTSs, are widely used in design process and system integration, adversaries may hide extra circuit(s) in them to undermine the system at a specific time or to gain control over it. The untrusted foundry issue is rooted in the outsourcing of design fabrication. Establishing a chip fabrication factory is extremely expensive and most semiconductor companies have become fabless in recent years. They ask foundries to fabricate their designs to reduce the overall cost. The third party, however, may change the designs by adding extra circuits, like back doors to receive confidential information from the chip, or altering circuit parameters, like wire thickness to cause a reliability problem in the field. The PCB assembly is even susceptible, as it is possible to mount extra components on interfaces between genuine components. In short, cooperative system development process creates opportunities for malicious parties to take control of the system and to run vicious activities. Therefore, as a part of the system development process, security features should be installed to facilitate trustworthiness, validation, and to unveil any deviation from genuine specifications.

1.1.1 Hardware Trojans

The practice of outsourcing design and fabrication in the interest of economy, has raised serious national security concerns, since an adversary can subvert a design by adding extra circuits, called hardware Trojans [1]. In general, a hardware Trojan is defined as any intentional alteration to a design in order to alter its characteristics. A hardware Trojan has a stealthy nature and can alter design functionality under rare conditions. It can serve as a time bomb and disable a system at a specific time, or it can leak secret information through side channel signals.

A Trojan may affect circuit AC parameters such as delay and power; it also can cause malfunction under rare conditions. Shown in Fig. 1.2, a hardware Trojan consists of Trojan payload and Trojan trigger [16]. A functional Trojan takes inputs

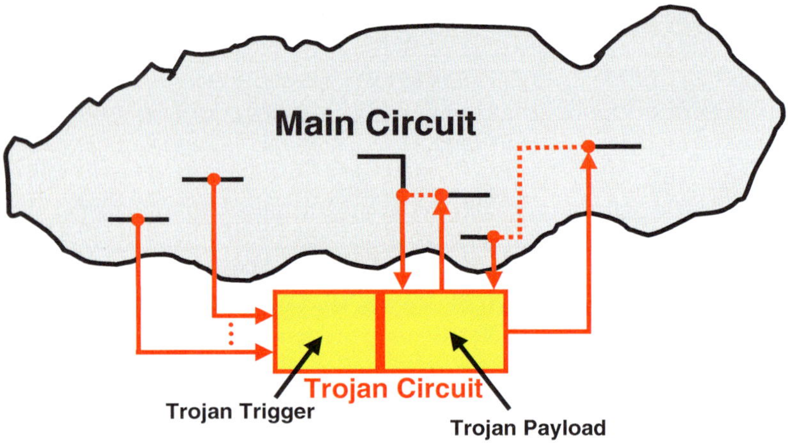

Fig. 1.2 Functional Trojan implementation

from some internal nets of the main circuit to the Trojan payload and restitches some other nets of the main circuit through Trojan payload to modify design functionality. The Trojan trigger determines the activation condition(s) under which the Trojan payload can propagate erroneous values into the main circuit.

Wang, Tehranipoor, and Plusquellic developed the first detailed taxonomy for hardware Trojans [7, 8]. This comprehensive taxonomy lets researchers examine their methods against different Trojan types. Currently, the industry lacks metrics to evaluate the effectiveness of methods in detecting Trojans. Such metrics could foster a comprehensive taxonomy to help analyze Trojan detection techniques. Because malicious alterations to a chip's structure and function can take many forms, Wang and colleagues decomposed the Trojan taxonomy into three main categories (see Fig. 1.3) according to their physical, activation, and action characteristics. Although Trojans could be hybrids of this classification (for instance, they could have more than one activation characteristic), this taxonomy captures the elemental characteristics of Trojans and is useful for defining and evaluating the capabilities of various detection strategies.

The physical characteristics category describes the various hardware manifestations of Trojans. The type category partitions Trojans into functional and parametric classes. The functional class includes Trojans that are physically realized through the addition or deletion of transistors or gates, whereas the parametric class refers to Trojans that are realized through modifications of existing wires and logic. The size category accounts for the number of components in the chip that have been added, deleted, or compromised. The distribution category describes the location of the Trojan in the chip's physical layout. The structure category refers to the case when an adversary is forced to regenerate the layout to insert a Trojan, which could then cause the chip's physical form factor to change. Such changes could result in different placement for some or all design components. Any malicious changes in physical layout that could change the chip's delay and power characteristics would

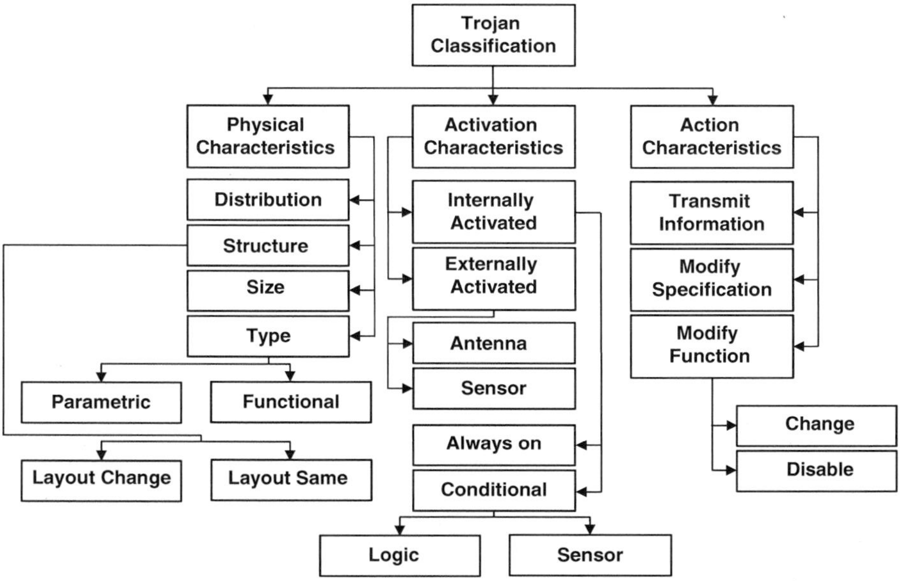

Fig. 1.3 Hardware Trojan Taxonomy

facilitate Trojan detection. Wang and colleagues identified current adversaries' capabilities for minimizing the probability of detection.

Activation characteristics refer to the criteria that cause a Trojan to become active and carry out its disruptive function. Trojan activation characteristics fall into two categories: externally activated (e.g., by an antenna or a sensor that can interact with the outside world) and internally activated (which are further classified as always on and condition based), as Fig. 1.3 shows. "Always on" means the Trojan is always active and can disrupt the chip's function at any time. This subclass covers Trojans that are implemented by modifying the chip's geometries such that certain nodes or paths have a higher susceptibility to failure. The adversary can insert the Trojans at nodes or paths that are rarely exercised. The condition-based subclass includes Trojans that are inactive until a specific condition is met. The activation condition could be based on the output of a sensor that monitors temperature, voltage, or any type of external environmental condition (such as electromagnetic interference, humidity, altitude, or temperature). Alternatively, this condition could be based on an internal logic state, a particular input pattern, or an internal counter value. The Trojan in these cases is implemented by adding logic gates and/or flipflops to the chip, and hence is represented as a combinational or sequential circuit.

Action characteristics identify the types of disruptive behavior introduced by the Trojan. The classification scheme shown in Fig. 1.3 partitions Trojan actions into three categories: modify function, modify specification, and transmit information. The modify-function class refers to Trojans that change the chip's function by adding logic or by removing or bypassing existing logic. The modify-specification class refers to Trojans that focus their attack on changing the chip's parametric properties, such as delay when an adversary modifies existing wire and transistor geometries. Finally, the transmit-information class includes Trojans that transmit key information to an adversary.

Trojan circuits are sly, triggering only under rare conditions. Trojans are designed to be silent most of their lifetime, to have a very small size relative to their host designs, and to make only limited contributions to circuit characteristics. Analyzing the vulnerabilities of IC development process requires the knowledge of design, fabrication, and test processes. To ensure a client's IC is authentic, the entire design and fabrication process must be made trustworthy or manufactured ICs should be verified by clients for trustworthiness. Having a separate and secure IC supply chain is desirable but economically prohibitive. Today, only Intel and few other companies still design and manufacture all their own chips in their own fabrication plants. Other chip designers have gone fabless, outsourcing their manufacturing to offshore facilities. In doing so, they avoid the huge expense of building a state-of-the-art fab, which, in 2007, cost as much as 2–4 billion in US dollars [1]. For example, the Petagon reports it now manufactures only 2 % of the more than $3.5 billion of integrated systems bought for military gears in secure facilities run by American companies [10]. These facts demand effective methods and techniques for Trojan prevention and detection.

1.1.1.1 Trojan Detection Methodologies

Several Trojan detection methodologies have been developed over the past few years. Without loss of generality, the methods are categorized as either side-channel analysis or Trojan activation, which are mainly chip-level solutions and architectural-level Trojan detection solutions.

Trojan Detection Using Side-Channel Signal Analysis

Side-channel signals, including timing and power, can be used for Trojan detection. Trojans typically change a design's parametric characteristics for example, by degrading performance, changing power characteristics, or introducing reliability problems in the chip. This influences power and/or delay characteristics of wires and gates in the affected circuit. Power-based side-channel signals provide visibility of the internal structure and activities within the IC, enabling detection of Trojans without fully activating them. Timing-based side channels can detect a Trojan's presence if the chip is tested using efficient delay tests that are sensitive to small changes in the circuit delay along the affected paths and that can effectively differentiate Trojans from process variations.

Power-Based Hardware Trojan Detection: In power-based techniques, the power consumption of IC under authentication (IUA) is compared with that of Trojan-free

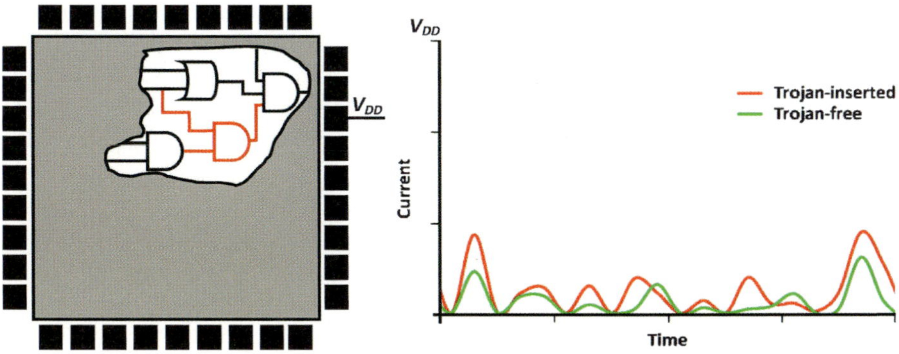

Fig. 1.4 Hardware Trojan detection based on power analysis

(golden) circuits. Figure 1.4 shows the measured current from VDD pin in Trojan-free and Trojan-inserted circuits over a specific time interval. Each current measurement consists of several elements, including (1) the main circuit current consumption which is the same for all chips, (2) measurement noise which can be eliminated by averaging several measurements, (3) process variations which are random and cannot be canceled, and, lastly, (4) Trojan contributions if they exist. Any measurable difference beyond process variations can be an indication of Trojan existence.

Agrawal et al. were the first to use side-channel information to detect Trojan contributions to circuit power consumption [3]. To obtain the power signature of Trojan-free (i.e., genuine) ICs, random patterns are applied and power measurement is performed. After patterns are applied, a limited number of ICs are reverse engineered to ensure they are Trojan free. Once the reference signature is obtained, the same random patterns are applied to the IC under authentication (IUA). If the IUA's power signature differs from the reference signature, the IUA is considered suspicious and it might contain a Trojan. Trojans of different sizes under different process variations are detected by applying random patterns and observing the signatures. If the Trojan is comparable in size with the circuit, its impact on the circuit-transient current will be significant and could be measured easily. However, process variations will mask the impact of very small Trojans on circuit power consumption.

Rad et al. proposed a region-based transient power signal analysis method to reduce the impact of increasing process variation levels and leakage currents [11]. A region is a portion of the layout that receives the majority of its power from surrounding power ports or C4 bumps. Measurements are made through each power port individually by applying patterns. The transient-current detection algorithm is based on a statistical analysis of the I_{DDT} waveform areas generated at each power ports as a test sequence is simulated on the design. For each orthogonal pairing of power ports, a scatter plot is constructed. The authors used several different process models for Trojan-free and Trojan-inserted designs. A prediction ellipse derived from a Trojan-free design with different process models can help distinguish between Trojan-inserted and Trojan-free designs. The dispersion in the Trojan-free data points is a result of uncalibrated process and test environment (PE) variations. However, regional analysis alone is not sufficient for dealing with the adverse effects of PE variations on detection resolution. Signal calibration techniques are necessary to attenuate and remove PE signal variation effects, to fully leverage the resolution enhancements available in a region-based approach. Calibration is performed on each power port and for each chip separately, and it measures the response of each power port to an impulse. After each test pattern is applied, the response is calibrated using the calibration matrix. The results presented by Rad et al. show that calibration can increase the distance between Trojan-free and Trojan-inserted designs under different process parameters.

Alkabani and Koushanfar proposed several approaches for gate-level timing and power characterization via nondestructive measurements [13]. Each measurement forms one equation. After a linear number of measurements are taken, a system of equations for mapping the measured characteristics to the gate level is formed. Potkonjak et al. exploited the formulation of gate-level characterization using linear programming and singular-value decomposition to detect Trojans [14]. They used both timing and static-power measurements. Trojan detection is performed via constraint (equation) manipulation. This method attempts to find the measurement matrix with the highest rank, and derives several heuristics for detecting gates that have inconsistent characteristics compared to their original specified characteristics. Learn, test, and resubstitution statistical validation techniques are used to estimate the bounds for normal (non-malicious) characteristics. The experiments considered errors in noninvasive measurements, but not process variations. The evaluation results are promising because gate-level characterization with high accuracy is possible. The gate-level characterization methods can find the characteristics of controllable gates. This controllability is known to be high for static power measurements and I_{DDQ} testing. Alkabani and Koushanfar used statistical convergence of gate-level estimation and signal integrity for Trojan detection [13]. They found efficient robust approximations for gate power consumptions and identified malicious insertions using multiple consistency checking.

Delay-Based Hardware Trojan Detection: There are also techniques that analyze the impact of Trojans on design performance. Any additional gates or wiring introduces extra capacitances, and then any rising or falling on Trojan-inserted paths creates extra time for transition. Figure 1.5 shows that the Output_Tx signal in a Trojan-free circuit changes sooner compared with a Trojan-inserted circuit. The signal over the highlighted path passes through extra wiring and an additional gate and experiences additional delay due to the resistance and capacitance of the extra wiring and transport delay of the Trojan gate.

In [6] a path delay fingerprint is proposed which is basically similar to [3] but based on analyzing circuit delay. A Trojan, even one small in size compared to the size of the main circuit, can have impact on at least one path. A circuit has many paths, each representing one part of the entire circuit characteristic.

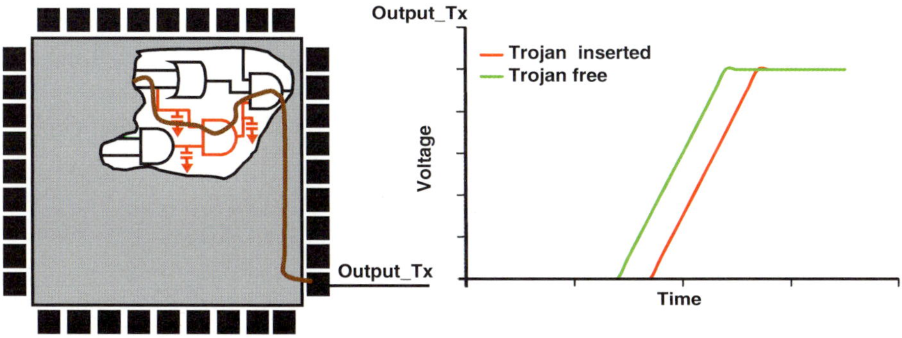

Fig. 1.5 Hardware Trojan detection based on delay analysis

The technique measures the delay of several nominated paths on several chips to bring process variations into account. Afterwards, the chips are reversed engineered to ensure they are genuine, and their measurements are used as a signature. The same measurements are performed on other chips and compared with the signature. Any difference can be an indication of Trojan insertion.

As another delay-based approach, a specific delay measurement circuit based on shadow registers is provided to measure the delay of a candidate path in [9]. The technique is mainly used for IC characterization and it can be utilized for hardware Trojan detection, as well. The delay of a nominated path between two registers (source and destination registers) is characterized by a shadow register which has a clock (clk2) with the same frequency as the clock applied to the registers (clk1), but with a negative phase shift (i.e. negative skew). To characterize the path, clk2 is applied with different skew till the captured data in the shadow register and destination register become different, and then clk2, along with the pattern applied to the path under test, is stored. Two measurements are performed at the design time and test time. The design-time measurement is performed on nominated paths with different process variations to develop a statistical data for each path. At the test time the same measurement is performed on each path and compared with the stored statistical data. Any significant difference between stored clk2 at design time and obtained clk2 at test time indicates Trojan existence.

Trojan Activation Methods

Trojan activation strategies can accelerate the Trojan detection process, and in some cases have been combined with power analysis during implementation. If a portion of the Trojan circuitry is activated, the Trojan circuit will consume more dynamic power, which will further help differentiate the power traces of Trojan-inserted and Trojan-free circuits. The existing Trojan activation schemes can be categorized as follows.

Region-Free Trojan Activation: These methods do not rely on the region but depend on accidental or systematic activation of Trojans. For example, Jha and Jha presented a randomization-based probabilistic approach to detect Trojans [15]. They showed that it's possible to construct a unique probabilistic signature of a circuit on the basis of a specific probability for patterns applied to its inputs. They apply input patterns based on the specific probability to IUA and compare its outputs with the original circuit. If there are differences in the outputs, a Trojan is present. For Trojan detection in a manufactured IC, patterns can be applied only on the basis of such probability to obtain a confidence level regarding whether the original design and the fabricated chip are the same. Wolff et al. analyzed rare-net combinations in designs [16]. These rarely activated nets are used as Trojan triggers. At the same time, nets with low observability are used as payloads. Wolff et al. generated a set of vectors to activate such nets and suggested combining them with traditional ATPG test vectors to activate a Trojan and to propagate its impact if the Trojan was connected to these nets.

Region-Aware Trojan Activation: Banga and Hsiao developed a two-stage test generation technique that targets magnifying the difference between the IUA and the genuine design power waveforms [4]. In the first stage (circuit partitioning), a region-aware pattern helps identify the potential Trojan insertion regions. To detect a Trojan circuit, the activity within a portion of the circuit is increased while the activity for the rest of the circuit is simultaneously minimized. The flip-flops in a circuit are classified into different groups, depending on structural connectivity. In the next stage (activity magnification), new test patterns concentrating on the identified regions are applied to magnify the disparity between the original and Trojan-inserted circuits. Regions (a set of flip-flops) exhibiting increased relative activity are identified by using the vector sequence generated in the first stage to compare the relative differences between the power profiles of the genuine and Trojan circuits. In this stage, more vectors for the specific regions, marked as possible Trojan regions, are generated using the same test generation approach as in the circuit-partitioning stage. Banga and Hsiao discussed magnifying Trojan contributions by minimizing circuit activity [5]. This involves keeping input pins unchanged for several clock cycles. Thus, circuit activity comes from the state elements of the design. Overall switching activity is therefore reduced, and can be limited to those specific portions of the design that help Trojan localization. Different portions of the design can be explored by changing input vectors to localize a Trojan. At the same time, each gate is equipped with two counters: TrojanCount and NonTrojanCount. With each vector, if the number of transitions at a gate's output exceeds a specific threshold, its TrojanCount would increase, and vice versa. The TrojanCount/NonTrojanCount ratio, called the gate weight, indicates a gate's activity. A high gate-weight ratio means the gate is considerably impacted by a Trojan, because there is a relatively high power difference corresponding to that gate's activation. Because the test engineer does not know the Trojan type or size, both region-free and region-aware methods are necessary. If a Trojan circuit's inputs come from the part of the circuit where they are functionally dependent (i.e., part of the same logic cone), the region-aware method can be effective. However, if the Trojan inputs are randomly selected from various parts of the circuit, region-free methods could increase the probability of detection.

Architecture-Level Trojan Detection

Verbauwhede and Schaumont explored trust issues at different levels of design abstraction (protocols, software, microarchitecture, and circuits) [17]. At the most abstract level, the adversary can access the interpreter and perform software tempering, scan-chain readout, or a fault attack. Side-channel information can be used at the software-architecture level. At the hardware microarchitecture and circuit levels, the attacker takes into account power energy consumption or electromagnetic energy. Hence, the authors proposed a systematic countermeasure to protect the root of trust at different design abstractions.

Tamper-proof techniques such as placing security parts into special casing with light, temperature, tampering, or motion sensors can provide protection at the physical level. Side-channel information such as power consumption should be separated from processing data or execution time to provide circuit-level protection. To deal with power fluctuation, different technologies such as full-custom dynamic and differential logic styles should be used. In experiments conducted by the authors, advanced encryption standards employing wave dynamic and differential logic remained safely after 1.5 million power-differential attack measurements, whereas standard CMOS technology disclosed the key only after 2,000 attack measurements.

To deal with side-channel attacks at the microarchitecture level, Verbauwhede and Schaumont suggested balancing if-and-else instructions to use the same amount of time and power during execution. The structure of microprocessors providing potential sources of side-channel information should be considered seriously. The authors also suggested using secure algorithm techniques, such as key and exponent blinding, to disable side-channel attacks at lower levels.

Suh, Deng, and Chan proposed authenticating the hardware by directly checking its implementation details at a low level [18]. The microarchitecture features of a high-end secure microprocessor are complex and unique for each model. A secure processor is authenticated by a checksum response to a challenge within a time limit. The unique checksum is based on the cycle-to-cycle activities of the processor's specific internal microarchitectural mechanism. Privacy is not breached, because the checksum depends on the processor-manufactured model and not the specific processor. The authors showed that small differences in the crypto-architecture result in significant deviations in the checksum. Their work relied on the speed advantages of the actual processor rather than simulations that attempt to impersonate the processor. The time limit on the authentication ensures resiliency against simulation models attempting to compute the checksum.

Bloom, Narahari, and Simha introduced a runtime Trojan activity detection mechanism using a hardware guard circuit and operating-system support [19].

Trojan attacks can either be internally or externally activated, and they can cause denial of service, privilege escalation, or leakage of sensitive information. Trojans can be detected by failure analysis and hardware verification, ATPG, or side-channel analysis. Bloom, Narahari, and Simha's work concentrated on denial-of-service (DoS) and privilege escalation attacks [19]. They used a hardware guard circuit to efficiently perform the testing, while the operating system generated the checks. Their hardware circuit included a timer, a scratch RAM, a simple processor, and an optional content-addressable memory (CAM).

Two tests were proposed: liveness checks and memory protection checks. Liveness checks are pseudo-random noncached-memory accesses that prevent simple prediction, delay, and replay attacks. Two solutions were provided for memory protection: a naive solution and a solution using a real-time operating system (RTOS). The naive solution periodically schedules a process that continuously tries to read the kernel memory. However, the process is time-consuming. RTOS support is needed to control the time of the checking process, which is created as a real-time task that is frequently required and consumes less time. The proposed solutions are evaluated on SPECit 2006 benchmarks. The overhead for using RTOS supportis approximately 2.2 %.

McIntyre et al. used hardware multicore systems, which permit simultaneous execution of the same functionality combined with verification [20]. Multicore systems are inherently redundant. Thus, as trust detection among the multiple cores is discovered, distributed software scheduling could be exploited to avoid low-trust cores. The distributed multicore task scheduler determines, over time and in the field, each core's hardware trust level.

Verifying the trustworthiness of manufactured ICs requires a post-manufacturing step to validate the conformance of the fabricated ICs to the original functional and performance specifications. Current design methodologies provide an adversary with multiple opportunities to insert Trojans that can go undetected. It is important to develop design-for-hardware-trust (DFHT) strategies (i) to prevent Trojan insertion into a design and (ii) to detect the Trojan if inserted. In other words, ICs must be designed in such a way that undetected changes to a circuit are near impossible.

1.1.2 Counterfeit ICs

Counterfeiting and piracy are longstanding problems which are growing in scope and magnitude. They are of great concern to governments because of (i) the negative impact they can have on innovation, (ii) the threat they pose to the welfare of consumers and (iii) the substantial resources that they channel to criminal networks, organized crime and other groups that disrupt and corrupt society. They are of concern to business because of the negative impact that they can have on (i) sales and licensing, (ii) brand value and firm reputation, and (iii) the ability of firms to benefit from the breakthroughs they make in developing new products [21].

Innovation in the business sector has always been the main driver of economic growth through the development and implementation of ideas for new products and processes. These inventions are usually protected via patents, copyrights, and trademarks. However, without adequate protection of these intellectual property (IP) rights, the incentives to develop these new ideas and products would be considerably reduced, thereby weakening the innovation process and critical thinking [21]. These risks are seen as particularly high for those industries in which the research and development (R&D) costs associated with the development of new products are very high compared to the cost of producing the resulting products. In the world of electronics, the R&D costs for the semiconductor industry are indeed extremely high, and protection of their IP rights is of utmost importance.

Without a doubt, counterfeiting of integrated circuits has become a major challenge due to the deficiencies in the existing test solutions. In the past couple of years, numerous reports (to be found in [22]) have pointed out to the counterfeiting issues in US electronic component supply chain. Senate Armed Services public hearing on this issue and the later report have clearly identified this as a major twenty-first century issue for US to address because of its significant implications on taxpayer money as well as the loss of lives that can be associated with deploying counterfeit parts in DOD critical applications [23, 24]. The report also indicated the lack of sufficient investment in this domain and that there are major shortcomings in detecting such counterfeit parts and the need to address them immediately.

In today's global economy, electronics components travel around the world before they make it into a system, such as, cell phone, computer, or security system. This global market has greatly reduced the cost of electronics, as large foundries can offer lower and lower prices. However, there is another illicit market willing to undercut the competition with equally illicit parts. If one of these ends up in consumer products, it will likely go undetected. The part may fail prematurely or unexpectedly, and the manufacturer will simply label the product as a defective unit and likely replace the product under warranty. However, if these parts end up in critical applications such as defense, aerospace, or medical, the results could be catastrophic. This is the market of counterfeits and it is stirring up serious problems in some sectors—including the United States Department of Defense [25].

Just how big the market is remains a mystery still. A study conducted from 2005–2007 [26] reveals that 50 % of original component manufacturers (OCM) and 55 % of distributors (authorized and unauthorized) have encountered counterfeit parts. The Electronic Resellers Association International [27] monitors, investigates, and reports issues that are affecting the global supply chain of electronics. ERAI, in combination with Information Handling Services Inc. [28], have been monitoring and reporting counterfeit component statistics dating back to 2001. The most recent data (Fig. 1.6) provided by IHS shows that reports of counterfeit parts have quadrupled since 2009.

Along with the increase of counterfeit incidents, it is also very important to analyze the vulnerabilities of the electronic components. Table 1.2 shows the five

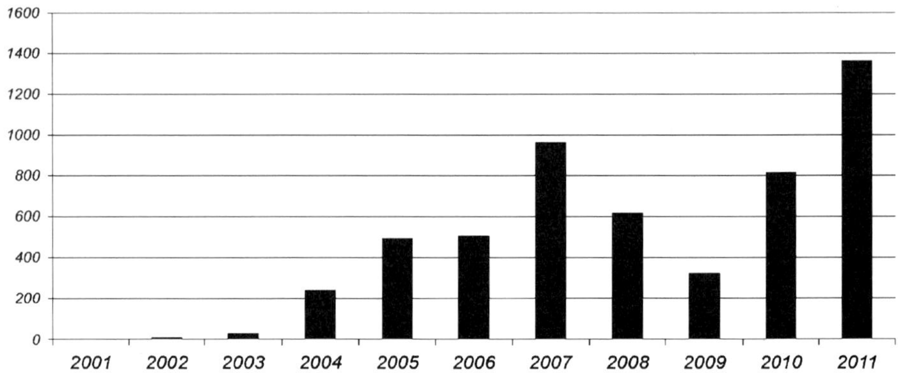

Fig. 1.6 Counterfeit incidents reported by IHS [29]

Table 1.2 Top-5 most counterfeited semiconductors in 2011 (Percentage of counterfeit part reports)

Rank	Commodity type	% of reported incidents (%)
#1	Analog IC	25.2
#2	Microprocessor IC	13.4
#3	Memory IC	13.1
#4	Programmable logic IC	8.3
#5	Transistor	7.6

Source: IHS Parts Management 2012 [30]

most commonly counterfeited components according to the percent of reported counterfeit incidents. They are as follows: analog ICs, microprocessor ICs, memory ICs, programmable logic ICs and transistors. Together, these five component groups contribute around 68 %, slightly more than two-thirds, of all counterfeit incidents reported in 2011. In this chapter, parts and components are used interchangeably to refer electronic devices.

This steady increase of reported incidents reflects the need for effective methods of testing parts and maintaining proper records as parts travel through the supply chain. There are a handful of available standards that seek to do just this, with more being written and revised constantly. The committee responsible for many of these standards is the G-19 Counterfeit Electronic Parts Committee, set forth by SAE International [31]. Their standards target three different sectors of the industry: distributors, users, and test service providers (i.e., test laboratories). A collection of the standards that they have written or are currently working on is as follows.

- AS6081—Counterfeit Electronic Parts Avoidance, **Distributors**
- ARP6178—Counterfeit Electronic Parts; Tool for Risk Assessment of Distributors, **Distributors & Users**
- AS5553—Counterfeit Electronic Parts; Avoidance, Detection, Mitigation, and Disposition, **Users**
- AS6171—Test Methods Standard; Counterfeit Electronic Parts, **Test Providers**

While SAE is the most prominent figure when it comes to standards and counterfeits, there are a couple of programs that are designed to help independent distributors gain trust from customers. Components Technology Institute, Inc. [32] is a multi-discipline company providing engineering and consulting services, training courses, and component conferences. They have created the Counterfeit Components Avoidance Program [33] (CCAP-101). Independent distributors can be certified as CCAP-101 compliant, which is done by means of a yearly audit. Another program with similar goals has been developed by the Independent Distributors of Electronics Association [34]. There is a comparison of the SAE's AS5553, CTI's CCAP-101, and IDEA's STD-1010 available in [35].

Note that these standards only deal with the detection of parts that are already in the market. There is another side to the anti-counterfeiting effort that takes on the prevention approach for parts that are currently being (will be) fabricated. Silicon physical unclonable functions (PUFs) have received much attention from the hardware security and cryptography communities as a new approach for IC identification, authentication and on-chip key generation [36–40]. Silicon PUFs exploit inherent physical variations (process variations) that exist in modern integrated circuits. These variations are uncontrollable and unpredictable, making PUFs suitable for IC identification and authentication [41, 42]. The variations can help generate a unique signature for each IC in a challenge-response form, which allows later identification of genuine ICs.

Due to the globalization of the semiconductor industry and the prohibitively high cost to create foundries and assembly companies for packaging, test, and burn-in processes, foundries now often fabricate the wafers/dies, test them and ship them to the assembly. The assembly then packages the dies, tests them, and ships the ICs to the market. The foundry/assembly however can ship defective, out-of-spec or even overproduced chips to the black market. The existing research on avoidance attempts to allow an IC designer to control the number of ICs produced. As an example, hardware metering approaches can be either passive or active. Passive approaches uniquely identify each IC and register the IC using challenge-response pairs. Later, suspect ICs taken from the market are checked for proper registration [37, 39, 43–46]. Active metering approaches, however, lock each IC until it is unlocked by the IP holder [42, 47–51]. This locking is done in a variety of ways including: (i) initializing ICs to a locked state on power-up [42], (ii) combinational locking by, for instance, scattering XOR gates randomly throughout the design [49–51], and (iii) adding a finite-state machine (FSM) which is initially locked and can be unlocked only with the correct sequence of primary inputs [48, 52].

Studying the vulnerabilities of electronic supply chain to counterfeiting is necessary to effectively address the problem. A comprehensive taxonomy of potential counterfeit component types reveals counterfeiters capability in forging. These shall shed light on challenges and foster efforts towards counterfeit detection and prevention.

References

1. S. Adee, "The Hunt for the Kill Switch," http://www.spectrum.ieee.org/print/6171.
2. A. Avizienis, J. Laprie, B. Randell, and C. Landwehr, "Basic Concepts and Taxonomy of Dependable and Secure Computing," IEEE Transactions on Dependable and Secure Computing, Vol. 1, No. 1, 2004, pp. 11–33.
3. D. Agrawal, S. Baktir, D. Karakoyunlu, P. Rohatgi and B. Sunar, "Trojan Detection using IC Fingerprinting," in Proc. of the *Symposium on Security and Privacy*, pp. 296–310, 2007.
4. M. Banga and M. S. Hsiao, "A Region based Approach for the Identification of Hardware Trojans," in Proc. of the *IEEE International Workshop on Hardware-Oriented Security and Trust (HOST2008)*, pp. 40–47, 2008.
5. M. Banga and M. S. Hsiao, "A Novel Sustained Vector Technique for the Detection of Hardware Trojans," in Proc. of the *International Conference on VLSI Design*, pp. 327–332, 2009.
6. Y. Jin and Y. Makris, "Hardware Trojan Detection using Path Delay Fingerprint," in Proc. of the *IEEE International Workshop on Hardware-Oriented Security and Trust (HOST 2008)*, pp. 51–57, 2008.
7. M. Tehranipoor and F. Koushanfar, "A Survey of Hardware Trojan Taxonomy and Detection," in *the IEEE Design & Test of Computers*, vol.27, no.1, pp. 10–25, 2010.
8. X. Wang, M. Tehranipoor, and J. Plusquellic, "Detecting Malicious Inclusions in Secure Hardware: Challenges and Solutions," in Proc. of the *IEEE International Workshop on Hardware-Oriented Security and Trust (HOST2008)*, pp. 15–19, 2008.
9. J. Li and J. Lach, "At-speed Delay Characterization for IC Authentication and Trojan Horse Detection," in Proc. of the *IEEE International Workshop on Hardware-Oriented Security and Trust (HOST08)*, pp. 8–14, 2008.
10. J. Markoff, "Old Trick Threatens the Newest Weapons," http://www.nytimes.com/2009/10/27/science/27trojan.html?_r=3&emc=etal
11. R. Rad, X. Wang, J. Plusquellic and M. Tehranipoor, "Power Supply Signal Calibration Techniques for Improving Detection Resolution to Hardware Trojans," in Proc. of the *International Conference on Computer-Aided Design (ICCAD08)*, pp. 632–639, 2008.
12. L. Wang, C. Wu, and N. Touba, "VLSI Test Principles and Architectures: Design for Testability," Morgan Kaufmann Publishers.
13. Y. Alkabani and F. Koushanfar, "Consistency-Based Characterization for IC Trojan Detection," in Proc. of the *International Conference on Computer-Aided Design (ICCAD09)*, pp. 123–127, 2009.
14. M. Potkonjak, A. Nahapetian, M. Nelson, and T. Massey, "Hardware Trojan Horse Detection Using Gate-Level Characterization," in Proc. of the *ACM/IEEE Design Automation Conference (DAC09)*, pp. 688–693, 2009.
15. S. Jha and S.K. Jha, "Randomization Based Probabilistic Approach to Detect Trojan Circuits," in Proc. of the *High Assurance Systems Engineering Symposium (HASE08)*, pp. 117–124, 2008.
16. F. Wolff, C. Papachristou, S. Bhunia, R.S. Chakraborty, "Towards Trojan Free Trusted ICs: Problem Analysis and Detection Scheme," in Proc. of the *Design, Automation and Test in Europe (DATE08)*, pp. 1362–1365, 2008.
17. I. Verbauwhede and P. Schaumont, "Design Methods for Security and Trust," in Proc. of the *Design, Automation and Test in Europe (DATE07)*, pp. 672–677, 2007.
18. G.E. Suh, D. Deng, and A. Chan, "Hardware Authentication Leveraging Performance Limits in Detailed Simulations and Emulations," in Proc. of the *ACM/IEEE Design Automation Conference (DAC09)*, pp. 682–687, 2009.
19. G. Bloom, B. Narahari, and R. Simha, "OS Support for Detecting Trojan Circuit Attacks," in Proc. of the *IEEE International Workshop on Hardware-Oriented Security and Trust (HOST09)*, pp. 100–103, 2009.

20. D. McIntyre, F. Wolff, C. Papachristou, S. Bhunia, D. Weyer, "Dynamic Evaluation of Hardware Trust," in Proc. of the *IEEE International Workshop on Hardware-Oriented Security and Trust (HOST09)*, pp. 108–111, 2009.
21. OECD (2007) http://www.oecd.org/dataoecd/13/12/38707619.pdf.
22. trust-HUB (2013) http://trust-hub.org/home.
23. U.S. Senate Committee on Armed Services (2012) Inquiry into Counterfeit Electronic Parts in the Department Of Defence Supply Chain http://www.armed-services.senate.gov/Publications/Counterfeit%20Electronic%20Parts.pdf.
24. U.S. Senate Committee on Armed Services (2012) Suspect Counterfeit Electronic Parts Can Be Found on Internet Purchasing Platforms http://www.gao.gov/assets/590/588736.pdf.
25. US Congress (2011) Ike Skelton National Defense Authorization Act for Fiscal Year 2011.
26. U.S. Department Of Commerce (2010) Defense Industrial Base Assessment: Counterfeit Electronics.
27. ERAI (2012) Electronic Resellers Association International (ERAI).
28. IHS (2012) Information Handling Services Inc.
29. IHS (2012) Reports of Counterfeit Parts Quadruple Since 2009, Challenging US Defence Industry and National Security.
30. IHS (2011) Top 5 Most Counterfeited Parts Represent a $169 Billion Potential Challenge for Global Semiconductor Market.
31. SAE (2012) SAE International.
32. CTI (2012) Components Technology Institute, Inc.
33. CTI (2011) Certification for Coutnerfeit Components Avoidance Program.
34. IDEA (2012) Independent Distributors of Electronics Association (IDEA).
35. CTI (2010) Comparison of AS 5553, CTI-CCAP-101B, and IDEA-STD-1010-A.
36. Pappu, R. (2001) Physical one-way functions.
37. Suh, G.E. and Devadas, S. (2007) Physical Unclonable Functions for Device Authentication and Secret Key Generation. Proc. of ACM/IEEE on Design Automation Conference: 9–14.
38. Kursawe, K. and Sadeghi, A.-R. and Schellekens, D. and Skoric, B. and Tuyls, P. (2009) Reconfigurable Physical Unclonable Functions - Enabling technology for tamper-resistant storage. Proc. of IEEE International Workshop on Hardware-Oriented Security and Trust: 22–29.
39. Kumar, S.S. and Guajardo, J. and Maes, R. and Schrijen, G.-J. and Tuyls, P. (2008) Extended abstract: The butterfly PUF protecting IP on every FPGA. Proc. of IEEE International Workshop on Hardware-Oriented Security and Trust: 67–70.
40. Bolotnyy, Leonid and Robins, Gabriel (2007) Physically Unclonable Function-Based Security and Privacy in RFID Systems. Proc. of IEEE International Conference on Pervasive Computing and Communications: 211–220.
41. Xiaoxiao Wang and Tehranipoor, M. (2010) Novel Physical Unclonable Function with process and environmental variations. Design, Automation Test in Europe Conference Exhibition: 1065–1070.
42. Alkabani, Yousra M. and Koushanfar, Farinaz (2007) Active hardware metering for intellectual property protection and security. 16th USENIX Security Symposium on USENIX Security Symposium: 20:1–20:16.
43. Lofstrom, K. and Daasch, W.R. and Taylor, D. (2000) IC identification circuit using device mismatch. Proc. of IEEE International Solid-State Circuits Conference: 372–373.
44. Lee, J.W. and Daihyun Lim and Gassend, B. and Suh, G.E. and van Dijk, M. and Devadas, S. (2004) A technique to build a secret key in integrated circuits for identification and authentication applications. Proc. of Digest of Technical Papers on VLSI Circuits: 176–179.
45. Su, Y. and Holleman, J. and Otis, B. (2007) A 1.6pJ/bit 96 % Stable Chip-ID Generating Circuit using Process Variations. Proc. of IEEE International on Solid-State Circuits Conference: 406–611.
46. Farinaz Koushanfar and Gang Qu and Miodrag Potkonjak (2001) Intellectual property metering. Information Hiding: 81–95.

47. Chakraborty, R.S. and Bhunia, S. (2008) Hardware protection and authentication through netlist level obfuscation. Proc. of IEEE/ACM International Conference on Computer-Aided Design: 674–677.
48. Alkabani, Yousra and Koushanfar, Farinaz and Potkonjak, Miodrag (2007) Remote activation of ICs for piracy prevention and digital right management. Proc. of IEEE/ACM international conference on Computer-aided design: 674–677.
49. Roy, J.A. and Koushanfar, F. and Markov, I.L. (2008) EPIC: Ending Piracy of Integrated Circuits. Proc. on Design, Automation and Test in Europe: 1069–1074.
50. Jiawei Huang and Lach, J. (2008) IC activation and user authentication for security-sensitive systems. Proc. of IEEE International Workshop on Hardware-Oriented Security and Trust: 76–80.
51. Baumgarten, A. and Tyagi, A. and Zambreno, J. (2010) Preventing IC Piracy Using Reconfigurable Logic Barriers. IEEE Design Test of Computers: 66–75.
52. Chakraborty, R.S. and Bhunia, S. (2009) HARPOON: An Obfuscation-Based SoC Design Methodology for Hardware Protection. IEEE Transactions on Computer-Aided Design of Integrated Circuits and Systems: 1493–1502.

Chapter 2
Hardware Trojan Detection: Untrusted Third-Party IP Cores

In general, third-party IP (3PIP) cores fall into one of three categories: soft, firm, and hard, depending on their format when they are supplied. Soft IP cores are described using VHDL or Verilog and are the most flexible and popular cores in practice. Firm cores are described and synthesized for specific libraries, while hard IP cores are described at the physical level and are supplied as layout or GDSII file. Since soft IP cores are the most widely used, detecting hardware Trojans in 3PIP, defined as IP trust, has gained significant attention in the recent years.

Hardware Trojans can be inserted into 3PIPs by IP vendors during IP design to steal security information/data from other IPs in the system. Detection of such Trojans is extremely difficult since there is no known golden model for 3PIPs as IP vendors usually provide specification and source code, both of which may contain Trojans. The conventional side-channel techniques for IC trust are not applicable to IP trust. When a Trojan exists in an IP core, all the fabricated ICs will contain Trojans. The only trusted component would be the specification from the SOC designer which defines the function, primary input and output, and other information about the 3PIP that they intend to use in their systems. A Trojan can be very well hidden during the normal functional operation of the 3PIP supplied as register transfer level (RTL) code. A large industrial-strength IP core can include thousands of lines of code. Identifying the few lines of RTL code in a soft IP core that represent a Trojan is an extremely challenging task.

A case study that tries to address the IP trust problem is presented in [1]. This chapter will present this case study in details. Several concepts such as formal verification, code coverage analysis, and ATPG methods are employed in this case study to achieve high confidence, whether the circuit is Trojan-free or Trojan-inserted. It is important to note that a 3PIP source code is largely Trojan free; only a few parts may be suspicious. The challenge is to identify the suspicious parts that are most likely to be part of a Trojan. Suspicious signals are identified first by coverage analysis. Removing redundant circuit and equivalence theorems reduces

M. Tehranipoor et al., *Integrated Circuit Authentication: Hardware Trojans and Counterfeit Detection*, DOI 10.1007/978-3-319-00816-5_2,

the number of suspicious signals. Sequential ATPG is used to generate patterns to activate these suspicious signals. This method considers both the characteristics of dormant Trojans and the redundant circuit.

2.1 A Case Study for Hardware Trojan Detection in Third-Party Digital IP Cores

2.1.1 Formal Verification and Coverage Analysis

One of the important concepts in this case study is formal verification, an algorithmic-based approach to logic verification that exhaustively proves the functional properties of a design [2]. It contains three types of verification methods that are not commonly used in the traditional verification, namely model checking, equivalence checking, and property checking. All functions in the specification are defined as properties. The specific corner cases in the test suite as they monitor particular objects in a 3PIP could also be represented by properties, such as worry cases, inter-block interfaces, and complex RTL structures. They can be represented as properties wherever the protocols may be misused, assumptions violated, or design intent incorrectly implemented. Formal verification uses property checking to check whether the IP satisfies those properties. With property checking, every corner of the design can be explored. For example, in benchmark RS232, there are two main functionalities in the specification: (1) transmitter, and (2) receiver. Figure 2.1 shows the waveform of the transmitter. Take the *start bit* as an example; with *Rst* ==*1'b1*, *clk* positive edge, and *xmitH* == *1'b1*, the output signal *Uart_xmit* will start to transmit *start bit* "0". This functionality is described using the SystemVerilog property shown in Fig. 2.2, and the corresponding assertion is defined simultaneously. The remaining items in the specification are also translated to properties during formal verification. Once all the functionalities in the specification are translated to properties, coverage metrics can help identify suspicious parts in the 3PIP under authentication. Those suspicious parts may be Trojans (or part of Trojans).

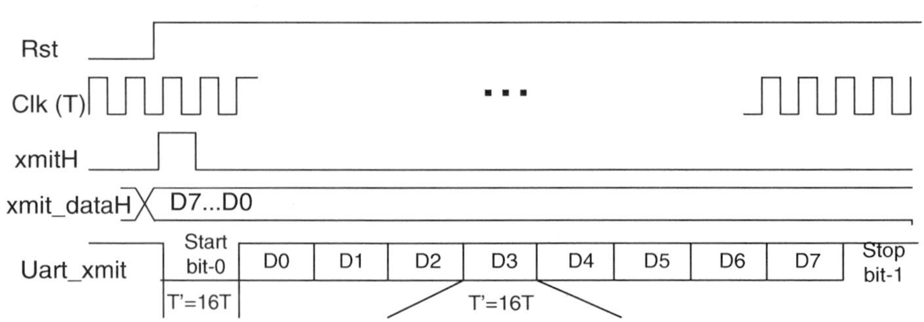

Fig. 2.1 Transmitter property in the specification stage

```
property e1;
  @(posedge uart_clk) disable iff (Rst)
  $rose(xmitH) |-> ##1 (uart_XMIT_dataH==0);
  endproperty

  a1: assert property( e1 );
```

Fig. 2.2 One of the properties and assertions definitions for RS232

```
Line No    Coverage    Block Type
  69           1        ALWAYS
  70           1        CASEITEM
  71           1        CASEITEM
  72           0        CASEITEM
  73           1        CASEITEM
  74           0        CASEITEM
  82           1        ALWAYS
  82.1         1        IF
...  ...      ...        ...
```

Fig. 2.3 Part of the line coverage report

Coverage metrics include code coverage and functional coverage. Code coverage analysis is a metric that evaluates the effectiveness of a test bench in exercising the design [3, 4]. There are many different types of code coverage analysis, but only a few of them are helpful for IP trust, namely line, statement, toggle, and finite state machine (FSM) coverage. Toggle coverage reports whether signals switch in the gate-level netlist while the other three coverage metrics show which line(s) and statement(s) are executed, and whether states in FSM are reached in RTL code during verification. Figure 2.3 shows parts of line coverage report during the simulation with RS232. This report shows that lines 72 and 74 are not executed, which helps improve the test bench by checking the source code. If the RTL code is easily readable, special patterns that can activate those lines will be added to the test bench. Otherwise, random patterns will be added to verify the 3PIP.

Functional coverage is the determination of how much functionality of the design has been exercised by the verification environment. The functional requirements are imposed on both the design inputs and outputs and on their interrelationships by the design specifications from SOC designer (i.e. IP buyers). All the functional requirements can be translated as different types of assertion, as in Fig. 2.2. Functional coverage checks those assertions to see whether they are successful or not. Table 2.1 shows part of the assertions coverage report (Assertion a1 is defined in Fig. 2.2). The number of Attempts in the table means there are 500, 003 positive edge clocks during the simulation time when the tool tries to check the assertion.

Table 2.1 Part of the assertion report with RS232

Assertion	Attempts	Real Success	Failure	Incomplete
test.uart1.uart_checker.a1	500,003	1,953	0	0
test.uart1.uart_checker.a2	1,953	1,952	0	1
...	...	...	...	...

The Real Success represents the assertion success rate while Failure/Incomplete denote the frequency of assertion failure/incomplete. When there are zero Failures, this property is always satisfied.

If all the assertions generated from the specification of the 3PIP are successful and all the coverage metrics such as line, statement, and FSM are 100 %, then it can be assumed with high confidence that the 3PIP is Trojan-free. Otherwise, the uncovered lines, statements, states in FSM, and signals are considered suspicious. All the suspicious parts constitute the suspicious list.

2.1.2 Techniques for Suspicious Signals Reduction

Based on the formal verification and coverage metric, a flow is proposed to verify the trustworthiness of 3PIP in [1]. The basic idea of the proposed solution is that without redundant circuit and Trojans in a 3PIP, all the signals/ components are expected to change their states during verification and 3PIP should function perfectly. Thus, the signals/components that stay stable during toggle coverage analysis are considered suspicious, as Trojan circuits do not change their states frequently. Each suspicious signal is then considered as the *TriggercEnablex*. Figure 2.4 shows the flow to identify and minimize the suspicious parts, including test pattern generation, suspicious signal identification, and suspicious signal analysis. Each step in the figure will be discussed in detail in the following.

2.1.2.1 Phase 1: Test Bench Generation and Suspicious Signal Identification

In order to verify the trustworthiness of 3PIPs, 100 % coverage of the test bench is best. However, it is very difficult to achieve 100 % coverage for every 3PIP, especially those with tens of thousands of lines of code. In the flow, the first step is to improve the test bench to obtain a higher code coverage with an acceptable simulation run time. With each property in the specification and basic functional test vectors, formal verification reports line, statement, and FSM coverage for the RTL code. If one of the assertions fails even once during verification, the 3PIP is considered untrustworthy, containing Trojans or bugs. If all the assertions are successful and the code coverage is 100 %, the 3PIP can be trusted. If at least one assertion fails or the code coverage is less than 100 %, more test vectors need to

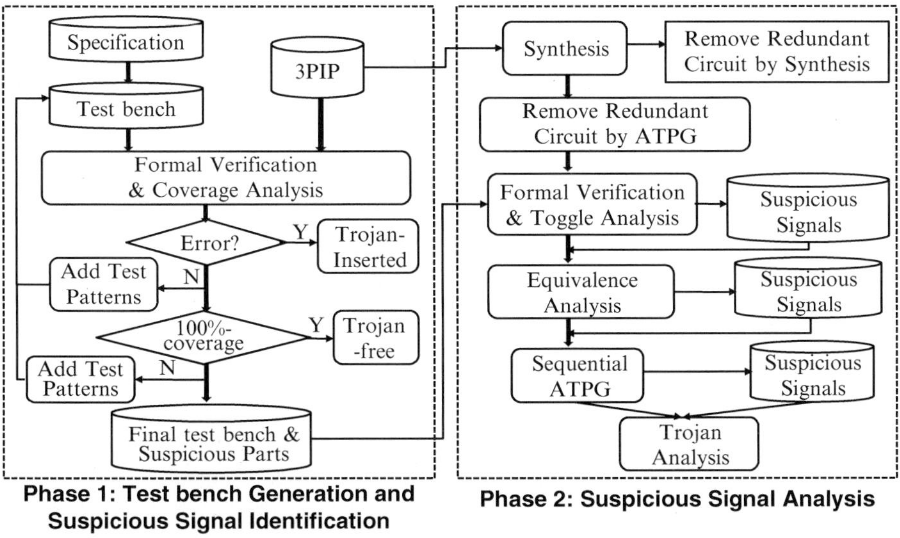

Fig. 2.4 The proposed flow for identifying and minimizing suspicious signals

be added to the test bench. The basic purpose of adding new vectors is to activate the uncovered parts as much as possible. But the verification time will increase as the number of test vectors increases. With the acceptable verification time and certain coverage percentage, both defined by the IP buyer, the final test bench will be generated and the RTL source code will be synthesized for further analysis.

2.1.2.2 Phase 2: Suspicious Signals Analysis

Redundant Circuit Removal (RCR): Redundant circuits must be removed from the suspicious list since they also tend to stay at the same logic value during verification, and input patterns cannot activate them. Removing a redundant circuit involves sequential reasoning, SAT-sweeping, conflict analysis, and data mining. The SAT method integrated in the Synopsys Design Compiler (DC) is used in this flow.

Another method to remove redundant circuits is developed in [1]. Scan chains are inserted into the gate-level netlist after synthesis for design testability, and ATPG generates patterns for all the stuck-at faults. The untestable stuck-at faults during ATPG are likely to be redundant logic. The reason is that if the stuck-at fault is untestable, the output responses of the faulty circuit will be identical to the output of the fault-free circuit for all possible input patterns. Thus, when ATPG identifies a stuck-at- 1/0 fault as untestable, the faulty net can be replaced by logic 1/0 in the gate-level netlist without scan-chain. All the circuits driving the faulty net will be removed as well. Figure 2.5a shows the circuit before redundant circuit removal.

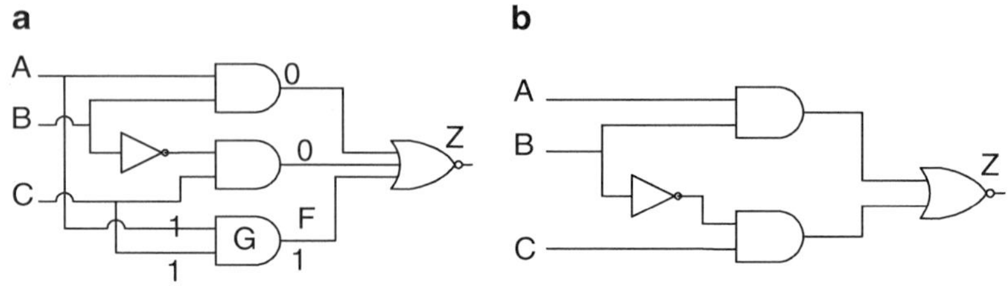

Fig. 2.5 (**a**) Before removing the redundant circuit with untestable F stuck-at-0 fault and (**b**) After removing the redundant circuit

The Stuck-at-0 fault of net F is untestable when generating patterns. Net F will be replaced by 0 and the gate G driving it will be removed from the original circuit, as shown in Fig. 2.5b.

After redundant circuit removal, toggle coverage analysis for the gate-level netlist without scan chain will identify which signals do not toggle (also called quiet signal) during verification with the test bench generated in Phase 1. These signals will be considered suspicious and added to the suspicious list. By monitoring these suspicious signals during verification, the authors obtain the logic value those signal are stuck at.

Equivalence Analysis: Fault equivalence theorems are known to reduce the number of faults during ATPG [5]. Similarly, the authors develop suspicious signal equivalence theorems to reduce the number of suspicious signals in [1].

Theorem 1. *If signal A is the D pin of a flip-flop (FF) while signal B is the Q pin of the same FF, the quiet signal A makes signal B quiet. Thus signal A is considered equal to B, which means if the pattern that can activate A is found, it will activate B as well. Then signal B will be removed from the suspicious signal list.*

As the QN port of a FF is the inversion of the Q port, they will stay quiet or switch at the same time. Thus the suspicious signal B would be considered equal to A and should be removed from the suspicious list.

Theorem 2. *If signal A is the output pin of an inverter while signal B is its input, they will stay quiet or switch at the same time. Thus the suspicious signal B would be considered equal to A and should be removed from the suspicious list.*

Theorem 3. *One of the input of AND gate A stuck-at-0 will cause the output B to stay quiet and one of the input of OR gate C stuck-at-1 will make the output D high all along. Thus, for AND gate, B stuck-at-0 is identical to A stuck-at-0, while for OR gate, D is identical to C stuck-at-1.*

Sequential ATPG: After reducing the number of suspicious signals by applying the above equivalence theorems, the authors use sequential ATPG to generate special patterns to change the value of certain signals during simulation in [1]. Stuck-at faults are targeted by the sequential ATPG to generate a sequential pattern to activate the suspicious signals when applied to the 3PIP. If the 3PIP functions perfectly with this pattern, the activated suspicious signals are considered part of the original circuit. Otherwise, there must be malicious inclusion in the 3PIP.

2.1.3 Simulation Results

The flow is applied to the RS232 circuit. 9 Trojans from the original design and 10 Trojans from [6] are inserted into the 3PIP. In total, there are 19 RS232 benchmarks with one Trojan in each IP. The following presents the simulation setup and test bench analysis for the 19 Trojan-inserted benchmarks. Next, the results of redundant circuit removal and the reduction of suspicious signals will be presented. Finally, Trojan coverage analysis will be discussed.

2.1.3.1 Benchmark Setup

Currently, there are over 80 benchmarks with different Trojans in the Trust-Hub [6], from which 51 benchmarks are at the RT Level. Readers can visit [6] for more details about the specification, structure, and functionality of these Trojans in the 10 RTL benchmarks. However, the other 9 Trojans are briefly described in the following:

*Trojan*1: The trigger of Trojan 1 is a special input sequence $8'ha6 - 8'h75 - 8'hc0 - 8'hff$. The payload changes the FSM in the transmitter of RS232 from state $Start$ to $Stop$, which means that once the Trojan is triggered, RS232 will stop transmitting data ($outputdata = 8'h0$). Since the trigger of the Trojan is a sequence of four special inputs, the probability of detecting the Trojan during verification is $1/2^{32}$. If the baud rate is 2,400 and RS232 transmits 240 words in one second, it will take 207.2 days to activate the Trojan and detect the error. In other words, it would be practically impossible to detect it by conventional verification. When this Trojan is inserted into RS232, an FSM is used to describe the Trojan input sequence. A three-bit variable state represents the FSM.

*Trojan*2: This Trojan only adds four lines to the original RTL code. If the transmitting word is odd and the receiving word is $8'haa$, RS232 will stop receiving words. This Trojan is less complex compared to Trojan 1, however, it provides opportunities to demonstrate the effectiveness of each step of the proposed flow.

*Trojan*3: The trigger of Trojan 3 is the same as that of Trojan 1, but the payload is different. Trojan 1 changes the state machine while Trojan 3 changes the shift process. The eighth bit of the transmitting word will be replaced by a Trojan bit during transmission. The Trojan bit could be authentication information, the special key to enable the system, or other important information.

*Trojan*4: Trojan 4 is designed to act like a time bomb. A counter is inserted into RS232 to count the number of words that have been sent out. After sending $10'h3ff$ words, the Trojan will be activated. The sixth bit of the transmitting word will be replaced by a Trojan bit.

Table 2.2 Analyzing the impact of the test bench on coverage metrics (a benchmark with Trojan 1 is used)

Test Bench #	Test Bench 1	Test Bench 2	Test Bench 3	Test Bench 4	Test Bench 5
Test patterns #	2,000	10,000	20,000	100,000	1,000,000
Verification time	1 min	6 min	11 min	56 min	10 h
Line coverage (%)	89.5	95.2	98.0	98.7	100
FSM state coverage (%)	87.5	87.5	93.75	93.75	100
FSM transition coverage (%)	86.2	89.65	93.1	96.5	100
Path coverage (%)	77.94	80.8	87.93	97.34	100
Assertion	Successful	Successful	Successful	Successful	Failure

*Trojan*5: After $24'hffffff$ positive edge clock, this Trojan's enable signal will become high. The sixth bit of the transmitting word will be replaced by a Trojan bit.

*Trojan*6: If RS232 receives "0" when the system is reset, the Trojan will be activated. The eighth bit of the transmitting word will be replaced by a Trojan bit.

*Trojan*7: When the transmitter sends a word $8'h01$ and the receiver receives a word $8'hef$ at the same time, the Trojan will be activated. A Trojan bit will replace the first bit of the transmitting word.

*Trojan*8 & 9: These Trojans do not tamper the original function of RS232 but add extra one stage (Trojan 8) and three stage (Trojan 9) ring oscillator to the RTL, which will increase the temperature of the chip quickly if they get activated.

2.1.3.2 Impact of Test Bench on Coverage Analysis

All the items in the specification are translated into properties and defined as assertions in the test bench. Assertion checkers will verify the correctness of assertions by SystemVerilog. Another important feature of a test bench is the input patterns. Some test corners need special input patterns. The more input patterns in the test bench, the more, for example, lines will be covered during verification. Table 2.2 shows five test benches with different test patterns and verification times for various coverage metric reports for the RS232 benchmark with Trojan 1. Generally, the verification time will increase with more test patterns and the code coverage will be higher as well. For Test Bench 1 to Test Bench 4, all the coverage reports are less than 100 % and all the assertions are successful, which indicates that the Trojan is dormant during the entire verification. The special test patterns added in Test Bench 5 increase the pattern count significantly and can activate the Trojans inserted in the benchmark. 100 % code coverage could be achieved with these additional test patterns. If one of the assertion experiences a failure, it signifies Trojan activation and the RS232 will give an erroneous output. One can conclude that the IP is Trojan-inserted. However, it is not easy to generate a test bench with 100 % code coverage for large IPs, and the verification time will be extremely long.

This phase of the flow can help improve the quality of the test bench. Given the time-coverage trade off, Test Bench 4 is selected for further analysis.

2.1.3.3 Reducing the Suspicious Signals

All the 19 benchmarks with different Trojans are synthesized to generate the gate-level netlist. The removal of redundant circuits is done during the synthesis process with special constrains using the Design Compiler. The simulation results are shown in Table 2.3. The second column in the table shows the area overhead of each Trojan after generating the final layout. As the table shows, Trojans are composed of different sizes, gates, and Structures, as well as different triggers and payloads as previously mentioned. The smallest Trojan has only 1.15 % area overhead. The percentage of Trojan area covered by suspicious signals *SS-Overlap-Trojan* is obtained by *SS-Overlap-Trojan*$=\frac{N_{SS}}{N_{TS}}$ where N_{SS} is the number of suspicious signals and N_{TS} is the number of Trojan signals. The results in Table 2.3 show that *SS-Overlap-Trojan* is between 67.7 % and 100 %, as shown in seventh column. If all the suspicious signals are part of the Trojan, the *SS-Overlap-Trojan* would be 100 %. This indicates that the number of signals in the final suspicious list fully overlapped with those from Trojan. This is an indicator of how successful the flow is at identifying Trojan signals. In addition, if the Trojan is removed or detected by sequential ATPG, the *SS-Overlap-Trojan* would also be 100 %.

Test Bench 4 is used to verify using the gate-level netlist and toggle coverage analysis reports which signals in each Trojan-inserted circuit are not covered by the simulation with all the successful assertions. Those quiet signals are identified as suspicious. The number of suspicious signals of each benchmark is shown in the third column of Table 2.3. Different benchmarks have a different number of suspicious signals based on the size of its Trojans. The larger the Trojan is, the more suspicious signals it has. On the other hand, the suspicious signals' stuck-at values are monitored by verification. All stuck-at-faults are simulated by the ATPG tool with scan chain in the netlist. If the fault is untestable, the suspicious circuit is a redundant circuit and will be removed from the original gate level netlist, in addition to the gates that drive the net. The number of suspicious nets after redundant circuit removal is shown in the fourth column of Table 2.3. As can be seen in the table, the suspicious nets of benchmarks with Trojan 8 and Trojan 9 are zero, which means that if the redundant circuits are removed in the two benchmarks, the benchmarks will be Trojan-free. The reason that redundant circuit removal can distinguish Trojans is that some Trojans are designed without payload and have no impact on circuit functionality. Thus it can be concluded that such Trojans can be removed by redundant circuit removal.

The remaining suspicious nets of each benchmark are needed to be processed by equivalence analysis and sequential ATPG. The fifth and sixth columns in Table 2.3 show the number of suspicious signals after the first two steps. It can be concluded that equivalence analysis can reduce a large number of suspicious signals, and sequential ATPG can be effective as well. For benchmarks with Trojan 2

Table 2.3 Suspicious signal analysis

Benchmark (RS232)	Trojan area overhead (%)	Step 1: Number of SS after RCR with Synthesis	Step 2: Number of SS after RCR with ATPG	Step 3: Number of SS after equivalence analysis	Step 4: Number of SS after sequential ATPG	SS-Overlap-Trojan (%)
With Trojan 1	11.18	22	20	17	12	100
With Trojan 2	20.35	17	16	3	Trojan is identified	100
With Trojan 3	10.48	20	15	15	10	97.3
With Trojan 4	20.35	3	3	3	2	87.6
With Trojan 5	4.59	9	8	8	7	100
With Trojan 6	1.15	1	1	1	Trojan is identified	100
With Trojan 7	3.79	3	3	3	2	100
With Trojan 8	1.15	1	Trojan is removed	-	-	100
With Trojan 9	3.79	3	Trojan is removed	-	-	100
TR04C13PI0	1.6	8	3	3	3	100
TR06C13PI0	1.8	9	3	3	3	100
TR0AS10PI0	2.09	8	1	1	1	100
TR0CS02PI0	25.3	59	55	39	39	67.7
TR0ES12PI0	2.09	8	1	1	1	100
TR0FS02PI0	25.0	30	28	20	20	73.3
TR2AS0API0	11.9	19	18	11	11	100
TR2ES0API0	12.0	20	18	11	11	100
TR30S0API0	12.4	22	20	13	13	93.6
TR30S0API1	12.3	25	22	14	14	87.3

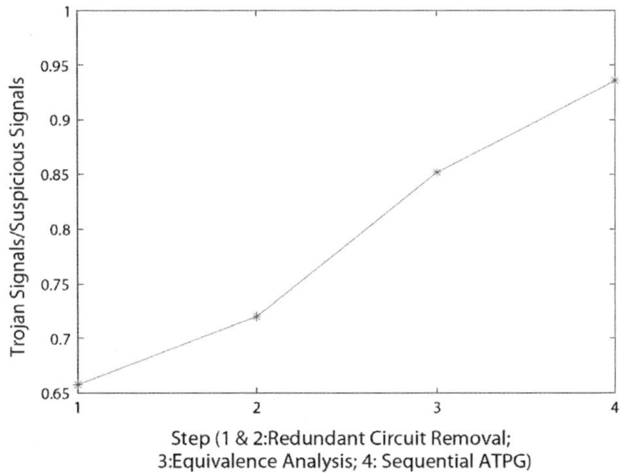

Fig. 2.6 Average Trojan signals/Suspicious signals in 19 benchmarks

and Trojan 6, the sequential ATPG can generate sequential patterns for the stuck-at faults in the suspicious signal. The sequential test patterns improve the test bench and increase its coverage percentage. Even though the coverage percentage is not 100 %, some assertions experience failure during simulation. Thus, the benchmarks with Trojan 2 and Trojan 6 are identified as Trojan-inserted.

The flow is implemented on 10 trust benchmarks from the Trust-Hub [6] and the results reported in rows 11–20 in Table 2.3 show that the presented flow can effectively reduce the total number of suspicious signals. In addition, as shown in seventh column, there is a good overlap between the number of suspicious signals and the actual Trojan signals inserted into each benchmark. However, some benchmarks experience low *SS-Overlap-Trojan*, such as RS232-TR0CS02PI0, since only part of this Trojan was activated during simulation.

2.1.3.4 Trojan Coverage Analysis

In the suspicious list, not all of signals are a result of Trojans. However, the $TriggerEnable$ signal must be in the suspicious list if the IP contains a Trojan. Once one net is identified as part a Trojan, it can be concluded that the 3PIP is Trojan-inserted. All the gates driving this net are considered to be Trojan gates. Figure 2.6 shows that the percentage of Trojan signals in the suspicious list increases significantly with the flow. As the authors apply different steps (step 1–4) to the benchmarks, 72 %, on average, of the suspicious signals are of the result of Trojans after redundant circuit removal with synthesis and ATPG in the 19 benchmarks. However, the percentage increases to 85.2 % when equivalence analysis is done and 93.6 % of signals in the suspicious signal list come from Trojans after sequential ATPG is applied to these benchmarks.

2.2 Summary

In this chapter, a case study is presented to verify the trustworthiness of 3PIPs, involving formal verification, coverage analysis, redundant circuit removal, sequential ATPG, and equivalence theorems. The code coverage generates the suspicious signals list. Redundant circuit are removed to reduce the number of suspicious signals. Equivalence theorems are developed for the same purpose. Sequential ATPG is used to activate these suspicious signals and some Trojans will be detected. However, more work is needed to get 100 % hardware Trojan detection rates in 3PIPs.

References

1. X. Zhang and M. Tehranipoor, "Case Study: Detecting Hardware Trojans in Third-Party Digital IP Cores," in Int. IEEE Hardware-Oriented Security and Trust (HOST), 2011.
2. A. J. Hu, "Formal Hardware Verification with BDDs: An Introduction," *IEEE*, 1997.
3. Synopsys, "The Synopsys Verification Avenue Technical Bulletin", Vol.4,issue 4, December 2004.
4. I. Ugarte and P. Sanchez, "Formal Meaning of Coverage Metrics in Simulation-Based Hardware Design Verification," IEEE, 2005.
5. M. Bushnell and A. Vishwani, "Essentials of Electronic Testing for Digital, Memory and Mixed-Signal VLSI Circuits," 2000.
6. http://trust-hub.org/resources/benchmarks.

Chapter 3
Hardware Trojan Detection: Untrusted Manufactured Integrated Circuits

Trojan circuits are designed to avoid detection, triggering only under rare conditions. Trojans are silent most of their lifetimes and have a very small size, relative to their host circuits, and make only limited contributions to circuit characteristics. These qualities suggest that they most likely connect to nets with low controllability and/or observability [1, 4].

Hardware Trojan detection depends on the Trojan's full or partial activation. In the full activation scenario, circuit functionality deviates from the genuine specifications, and the Trojan can cause catastrophic failures. In the partial activation scenario, however, the Trojan impacts the circuit power profile or its delay characteristics. Several techniques have been proposed to address these rare triggering conditions or to capture the impact of a Trojan on side-channel signals.

For full activation of a Trojan, in [2] circuit nets with low probabilities of "1" or "0" are distinguished. To avoid the Trojan's detection, an adversary may utilize a combination of low transition nets as a Trojan trigger to reduce the probability of activation during authentication. In this work, patterns are generated to make those nets more switch in order to increase the probability of Trojan activation. Many techniques have been also proposed based on Trojan partial activation by studying their impact on the delay or the power characteristics of circuit under authentication (CUA).

3.1 A Case Study for Hardware Trojan Detection in Integrated Circuits

The amount of current drawn by a Trojan can be so small that it submerges into the envelope of noise and process variation effects, where it cannot be detected by measurement equipment. However, Trojan detection probability can improve greatly when the current is measured locally from multiple power pads. The local current refers to the current drawn from a power port near a Trojan circuitry. The more

M. Tehranipoor et al., *Integrated Circuit Authentication: Hardware Trojans and Counterfeit Detection*, DOI 10.1007/978-3-319-00816-5_3,

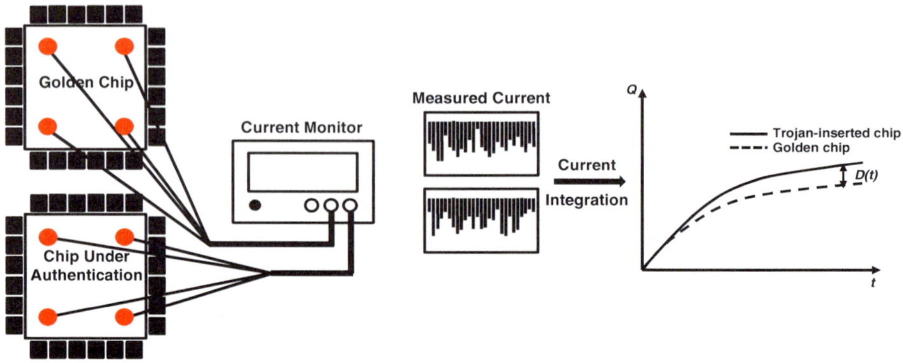

Fig. 3.1 Current integration method

instances of switching on Trojan inputs and inside Trojan circuitry, the greater the Trojan's power consumption. Since small Trojans are expected to be inserted into chips to reduce the probability of detection, the local current impact could be more significant than the global current measured from power pins.

Any partial activity in a Trojan circuit demands current. On the other hand, variations in process parameters, such as gate channel length and voltage threshold, sometimes increase or decrease the amount of circuit current consumption with different input vectors over time. Based on this fact, a current integration methodology is presented in [3] which accumulates the impact of a Trojan over the time while it is expected that the process variations impact is canceled out by integration. Figure 3.1 shows the current integration methodology for detecting hardware Trojans.

It is assumed that an adversary inserts Trojans randomly into a selected number of chips. Using an exhaustive test on a few randomly selected chips can help identify some golden chips. After identifying the golden chips, an average current waveform is formed in response to a pattern set. Next, the pattern set is applied to each CUA, and the current is measured locally via power pads or C4 bumps. Using this current integration method, the small current consumption difference between Trojan-inserted and Trojan-free circuits can be increased through the integration process. In the case of a Trojan's existence in a chip, more current difference can be measured by applying more patterns to the chip, making the Trojan detection task easier. When the current difference surpasses a pre-defined threshold, it indicates that Trojan detection and pattern application has stopped.

If $I_{Trojan-free}$ and $I_{Trojan-inserted}$ denote current drawn by Trojan-free and Trojan-inserted circuits at time t, respectively, the integrated current at time t for Trojan-free and Trojan-inserted circuits ($Q_{Trojan-free}(t)$ and $Q_{Trojan-inserted}(t)$) can be expressed by (3.1) and (3.2):

$$Q_{Trojan-free}(t) = \int_0^t I_{Trojan-free}(t).\mathrm{d}t \tag{3.1}$$

Table 3.1 Trojan characterization

Trojan	Type	Size (%)	Distribution	Structure
Counter	1-bit	0.04	Tight	No-change
	3-bit	0.10	Tight	No-change
	7-bit	0.31	Tight	No-change
	9-bit	0.42	Tight	No-change
Comparator	3-input	0.02	Loose	No-change
	5-input	0.04	Loose	No-change
	20-input	0.15	Loose	No-change

$$Q_{Trojan-inserted}(t) = \int_0^t I_{Trojan-inserted}(t).\mathrm{d}t = \int_0^t I_{Trojan-free}(t) + I_{Trojan}(t).\mathrm{d}t \qquad (3.2)$$

where $I_{Trojan}(t)$ denotes the current drawn by the Trojan. Since the same pattern set is applied to both a golden chip and a CUA, the difference between $I_{Trojan-free}$ and $I_{Trojan-inserted}(t)$ comes from (I) the additional current drawn by Trojan gates and (II) changes in the circuit current due to process variations. By integrating the current along the time axis for both chips, their cumulative difference at time t, denoted by $D(t)$ in (3.3), can be increased by applying more patterns.

$$D(t) = Q_{Trojan-inserted}(t) - Q_{Trojan-free}(t) = \int_0^t I_{Trojan}(t).\mathrm{d}t \qquad (3.3)$$

When $D(t)$ reaches a predefined Trojan detection threshold D_{th}, i.e. $D(t) \geq D_{th}$, then the chip is identified as a Trojan-inserted chip. It should be noted that D_{th} is determined by the Trojan detection timing budget as well as the current measurement device resolution.

The proposed current integration technique is effective for detecting both tightly- and loosely-distributed Trojans. Further, its capability does not depend on the location of a Trojan in a circuit, since current is measured locally through power pads or C4s. Since there are a large number of considerable power pads on the power distribution network, a Trojan circuitry impacts at least one power pad.

The current integration technique is used to detect Trojans inserted into the s38417 benchmark. First seven layouts of the original s38417 benchmark are generated using Synopsys physical design tools in the 180nm technology node. A 1-bit, 3-bit, 7-bit, 9-bit counter and 3-input, 5-input, 20-input comparator Trojan is inserted into each of these seven layouts (i.e. only one Trojan in each layout). Table 3.1 shows the type, size, distribution, and structure of the Trojans. The impact of variations in voltage threshold (V_{th}), channel length (L), and oxide thickness (T_{ox}) on Trojan detection is investigated as well. Table 3.2 shows the applied process variations.

Table 3.2 Process variations applied during trojan detection

Parameter	Inter-die (%)	Intra-die (%)
Threshold voltage (V_{th})	5	20
Channel length (L)	2	8
Oxide thickness (T_{ox})	1	4

Figure 3.2a shows the simulation results obtained using Synopsys Nanosim for s38417 containing the Trojan 9-bit counter. The patterns are applied with a frequency of 100 MHz. As seen in the figure, after applying 700 clock cycles (7us pattern application time), $D(t) > 1 \times 10^{-9}$, which is easily detectable using measurement devices. In fact, for such a Trojan, depending on D_{th}, a shorter application time can be even sufficient for detection. Shown in Fig. 3.2b, the results obtained for s38417 with a 7-bit counter also confirm that such a Trojan can be easily detected. In general, detecting a counter is easier than a combinational Trojan since a counter continuously receives the clock and consumes power. No process variations are considered for the results shown in Fig. 3.2, although the process variations would not be significant enough to change the detection outcome for such Trojans.

Figure 3.3a,b show the simulation results for the circuit with 3-bit and 1-bit counters, respectively, without considering process variations. The results imply that the smaller Trojans consume significantly lower power (i.e. current) which makes their detection more difficult. Note that Trojan detection depends on two important factors: (1) process variations and (2) the resolution of measurement device. It is believed that process variations are a limiting factor in Trojan detection.

Figure 3.4 shows the simulation results for the circuit containing a 3-bit counter considering the worst case process corners for both Trojan-inserted and Trojan-free circuits. The process corner used in the Trojan-free circuit increases the current within the circuit while the process corner used in the Trojan-inserted circuit reduces the total current. This is done to evaluate the efficiency of the technique in detecting the Trojan. The Trojan-inserted circuit with process variations still consumes more current compared with the Trojan-free circuit with process variations. However, detecting smaller Trojans, such as 2-bit and 1-bit counters, is not possible considering worst-case process variations. As seen in Fig. 3.3b, the current difference between Trojan-free and Trojan-inserted circuits containing 1-bit counters is negligible. The existence of process variations makes Trojan detection even more difficult.

Figure 3.5 shows simulation results for the 20-input comparator Trojan inserted in the s38417 benchmark after applying 300 random patterns. As seen in the figure, the 20-input comparator can be easily detected, even in presence of the two process corners. However, the results shown in Fig. 3.6 demonstrate the difficulty of detecting the 5-input comparator even without process variations. This is the case for the 3-input comparator Trojan as well. To further increase the probability of detection, more test patterns should be applied. The application time depends upon where the Trojan-inserted circuit's results fall outside of the results of Trojan-free

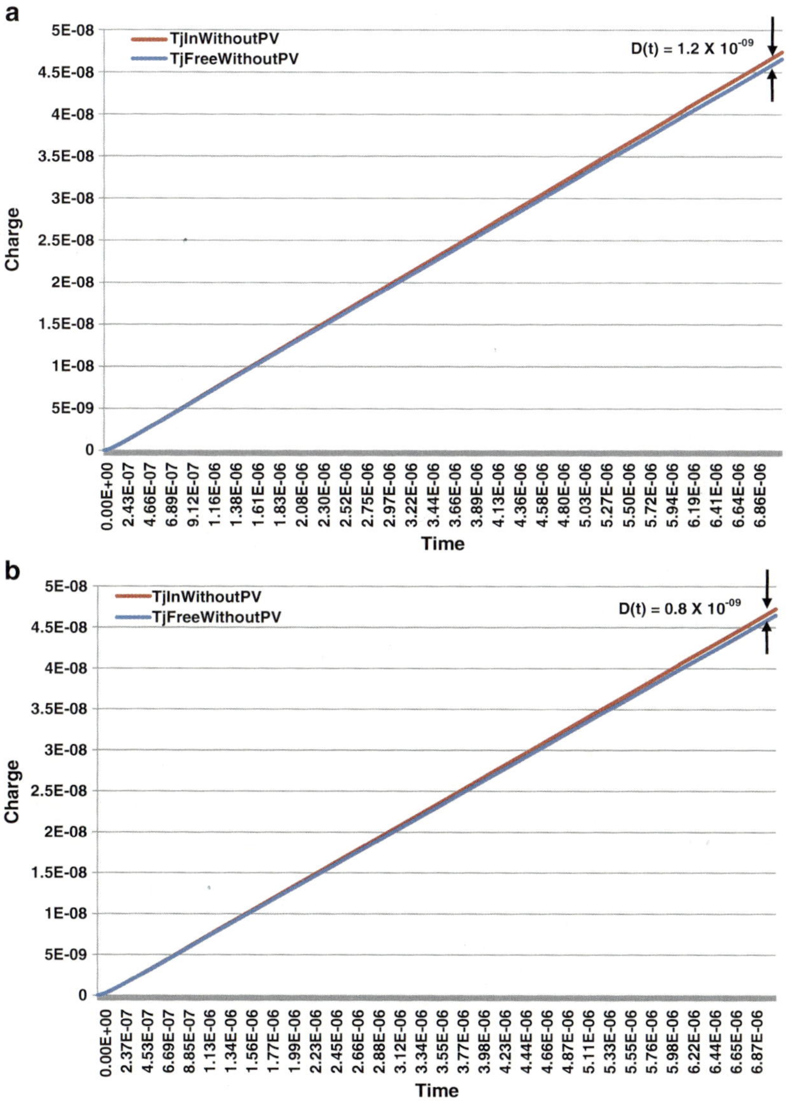

Fig. 3.2 Current integration for s38417 with 9- and 7-bit counters. (**a**) 9-bit counter. (**b**) 7-bit counter

circuit while considering process variations. The total number of patterns required to detect such small Trojans can be estimated from the results shown in Fig. 3.6 based on D_{th}.

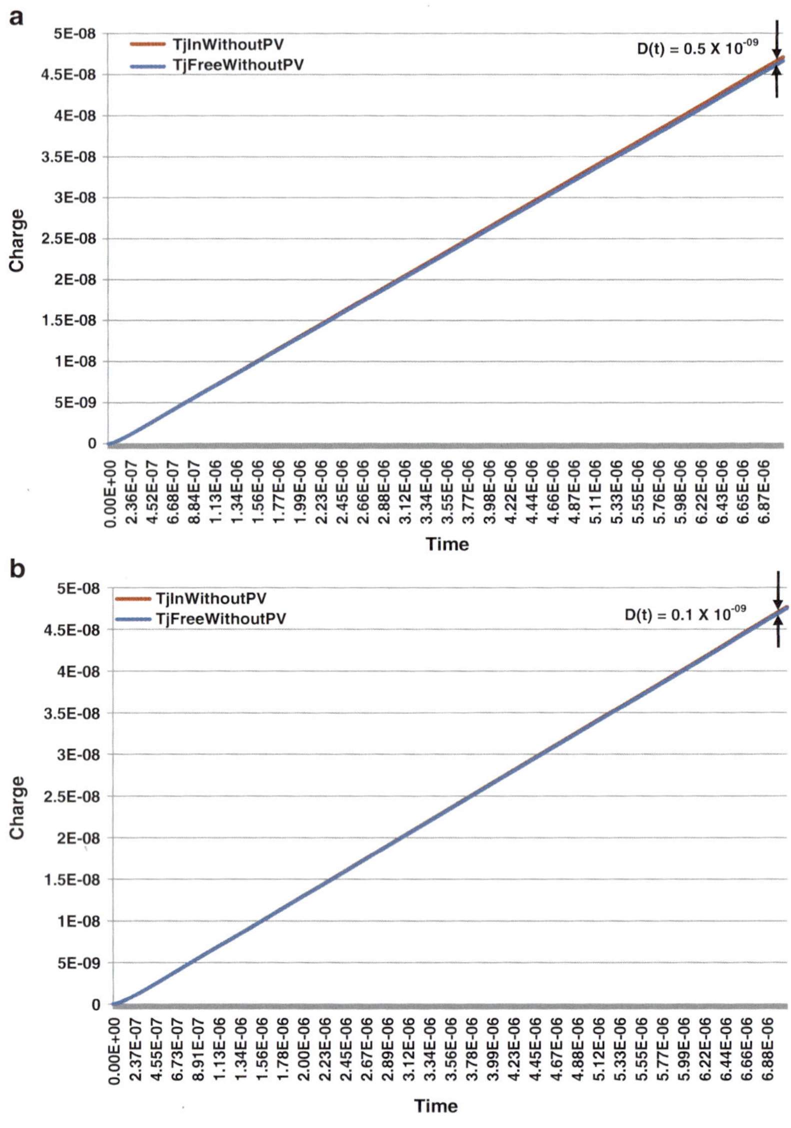

Fig. 3.3 Current integration for s38417 with 3- and 1-bit counters. (**a**) 3-bit counter. (**b**) 1-bit counter

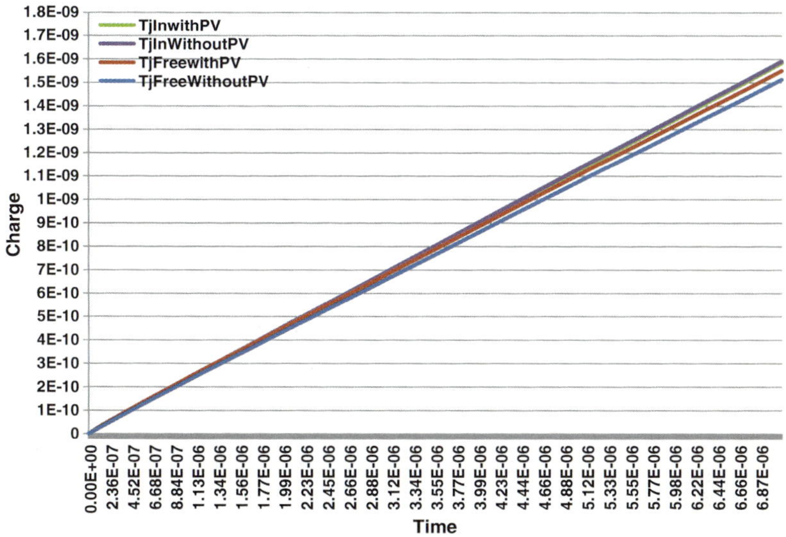

Fig. 3.4 Current integration for s38417 with 3-bit counter considering the two process corners

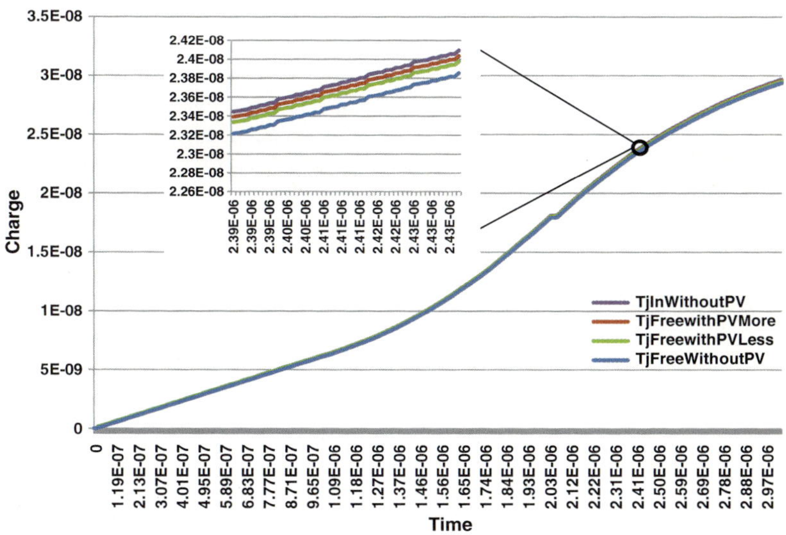

Fig. 3.5 Current integration for s38417 with 20-input comparator and two process corners

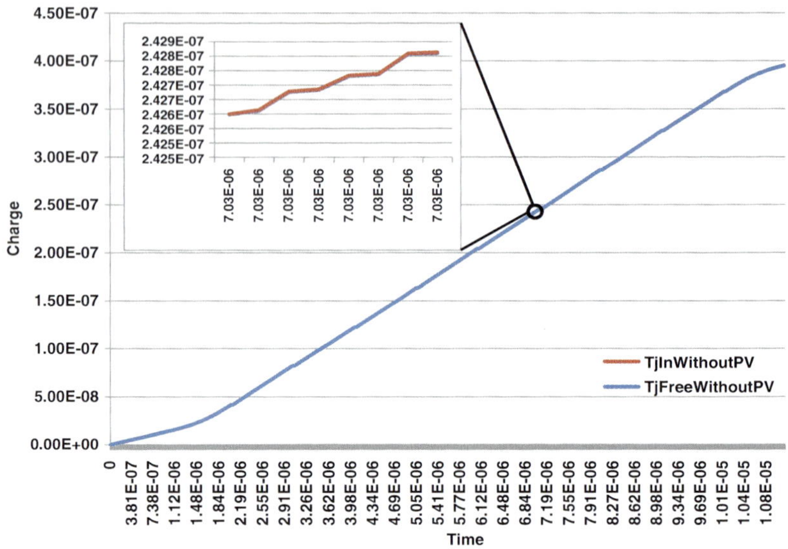

Fig. 3.6 Current integration for s38417 with 5-input comparator and no process variations

3.2 Summary

The current integration technique for Trojan detection and isolation was presented. The technique measures the current locally from the on-die power pads. Comparing the results obtained for a golden chip against a CUA, it can be seen that Trojans can be detected if the current integration results for the CUA fall outside that of the golden chip. It was shown that the technique can easily detect Trojans as small as 0.1 % within the circuit area.

References

1. M. Banga and M. S. Hsiao, "A Novel Sustained Vector Technique for the Detection of Hardware Trojans," in Proc. of the *International Conference on VLSI Design*, pp. 327–332, 2009.
2. F. Wolff, C. Papachristou, S. Bhunia and R.S. Chakraborty, "Towards Trojan-Free Trusted ICs: Problem Analysis and Detection Scheme," in Proc. of the *Design, Automation and Test in Europe(DATE '08)*, pp. 1362–1365, 2008.
3. X. Wang, H. Salmani, M. Tehranipoor and J. Plusquellic, "Hardware Trojan Detection and Isolation Using Current Integration and Localized Current Analysis," in Proc. of the *International Symposium on Fault and Defect Tolerance in VLSI Systems (DFT08)*, pp. 87–95, 2008.
4. X. Wang, M. Tehranipoor and J. Plusquellic, "Detecting Malicious Inclusions in Secure Hardware: Challenges and Solutions," in Proc. of the *IEEE International Workshop on Hardware-Oriented Security and Trust (HOST 2008)*, pp. 15–19, 2008.

Chapter 4
Design for Hardware Trust: Dummy Scan Flip-Flop Insertion

Hardware Trojan detection is an extremely challenging problem, and traditional structural and functional tests cannot effectively address it. Trojan inputs are supplied by nets with low-transition probabilities to reduce their impact on circuit side-channel signals such as power and delay. Automatic Test Pattern Generation (ATPG) methods used in manufacturing tests for detecting defects do so by operating on Trojan-free netlists. Therefore, existing ATPG algorithms cannot directly target Trojans [7].

Efficient pattern generation is necessary to disclose a Trojan's impact on circuit characteristics in the presence of process and environmental variations. Trojan detection methods based on transient power analysis [3–5, 8, 9] require patterns that increase Trojan activity while keeping circuit activity low to magnify the Trojan's contribution to the circuit power consumption. Methods based on delay analysis [1, 2] need patterns that generate transition on the nets that supply Trojan inputs in order to reveal wiring and input gate resistance and the Trojan's capacity to impact circuit delay characteristics. From an authentication standpoint, it is critical to (i) analyze the time required to generate a transition at a Trojan's input and within a Trojan circuit and (ii) to reduce authentication time.

4.1 Trojan Activation Time Analysis

Although, presumably, there is no information about a Trojan circuit in terms of size, type, or location, it is still crucial to analyze Trojan full and partial activation times. In *Full activation*, a Trojan becomes fully operational and causes malfunction. *Partial activation* is the generation of some transitions inside a Trojan circuit so that it improves the effectiveness of transient power-based methods [8, 9, 12]. In general, a functional Trojan consists of two parts: the Trigger and the Payload [13]. The Trojan Trigger is mostly inactive in nature, with no Payload effect. Under certain rare conditions or events, the Trojan becomes activated (triggered) and then the Payload injects an error to the main circuit. The ability to generate a transition in

M. Tehranipoor et al., *Integrated Circuit Authentication: Hardware Trojans and Counterfeit Detection*, DOI 10.1007/978-3-319-00816-5_4,

a Trojan circuit depends on its implementation. The switching of gates at the first level of the Trojan circuit depends upon their preceding cells.

In general, transitions in a circuit are induced by transitions in scan cells and at primary inputs [14]. A Trojan cone is defined as a logic circuit connecting to the inputs of a Trojan gate [15]. The structure of a Trojan cone, and the number and types of its gates can determine the time required to generate a transition in a Trojan gate. Figure 4.1 shows two example Trojan cones. These Trojans are named as *Trojan 1* and *Trojan 2*. *Trojan 1* contains three gates at two levels while *Trojan 2* contains seven gates at three levels. T_{g1} in *Trojan* 1 is connected to the cone shown in Fig. 4.1a and T_{g3} is connected to the cone in Fig. 4.1b. The other gates inside these two Trojans are assumed to be connected to other parts of the circuit.

In Fig. 4.1a, Trojan 1's cone consists of 17 gates at 11 levels. A Trojan cone contains all gates in the main circuit that impact with a Trojan gate, as well as the Trojan gate itself (here T_{g1}). Simulation results show that after applying 1,000 random test vectors in test-per-clock fashion, there are 67 transitions at the output of T_{g1}.

In Fig. 4.1b, Trojan 2's cone consists of 7 gates in 2 levels. The simulation results show that there are 421 transitions at the output of T_{g3} after applying 1,000 random test vectors. Since random vectors are applied to the above circuits, the results may be slightly different from one random vector set to another. As seen in the simulation results of Trojan 1 and Trojan 2, the number of transitions in the two Trojan gates varies significantly. This is mainly due to the difference in these Trojan cones' structures, number of levels, number of inputs (scan flip-flops and primary inputs), and the Trojan gate types.

Probability can represent circuit characteristics since it considers the gates' functionality and the interconnections between them. Knowing the probability of a node's switching in the circuit provides a good estimation of the time required to generate switching at the node. Suppose the probabilities of having "1" and "0" at the output of a Trojan are $P1$ and $P0$, respectively. In this case, the probability of switching from "0" to "1" or "1" to "0" at the output of the Trojan is $Pt_{Tgi} = P1 \times P0$, where T_{gi} is the ith gate at the first level of a Trojan. The probability of generating a transition at the output of Trojan gate T_{g3} (Pt_{Tg3}) is 0.25 by applying random patterns at the inputs of the Trojan 2' cone, as shown in Fig. 4.2. The circuit shown in this figure is the same as the one depicted in Fig. 4.1b.

To obtain the transition probability, a transition (i.e. success) can be modeled using Geometric Distribution (GD) [10]. The Geometric Distribution is a discrete distribution for $n = 0, 1, 2, \cdots$ with the probability function $p(n) = P \times (1 - P)^n$. The probability function states that after n clock cycles, finally in the $(n+1)$th clock cycle, there is a transition. In other words, the $(n+1)$th trial is the first success. The average number of experiments is $(P^{-1} - 1)$ which indicates the number of required clock cycles, *on average*, to generate a transition.

For the Trojan gate shown in Fig. 4.2, the calculation based on the Geometric Distribution shows that about three clock cycles are required to generate a transition at the output of Trojan gate (T_{g3}). The simulation results after applying 1,000 test vectors confirm that there is one transition every 2.37 clock cycles, on average.

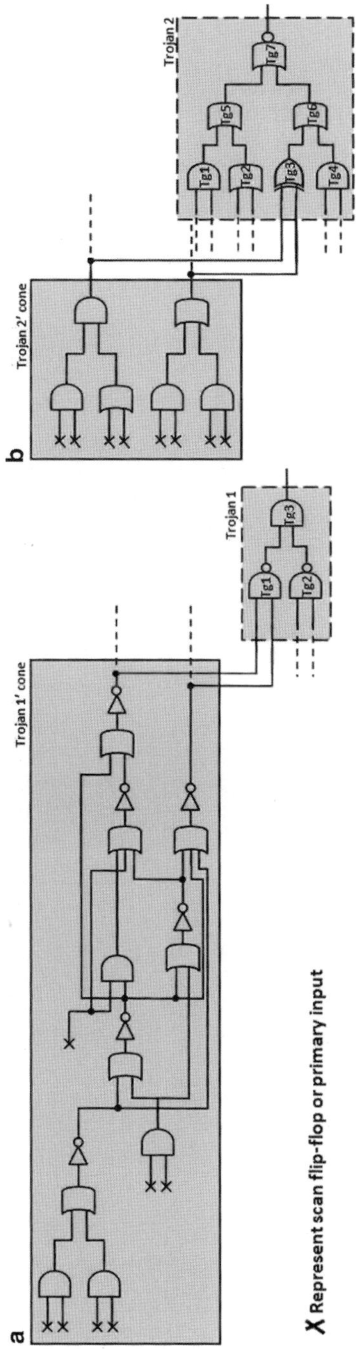

Fig. 4.1 Two Trojan cone examples: (**a**) Trojans 1 and (**b**) Trojan 2

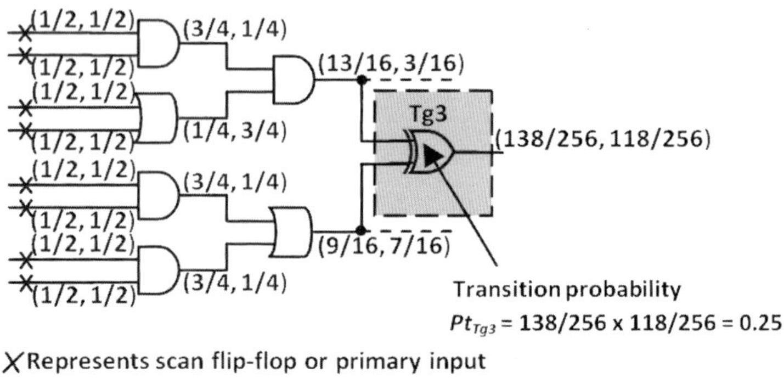

Fig. 4.2 Transition probability for a target cone

Fig. 4.3 Comparing mathematical and simulation results

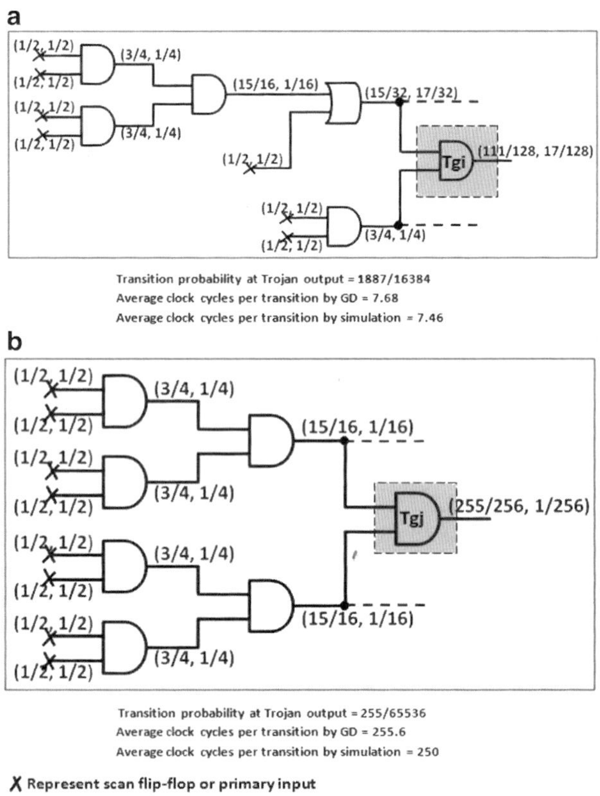

Figure 4.3 presents two new Trojan cones and compares the average clock cycles per transition using GD (i.e. probability analysis) and in simulation. Figure 4.3a shows that the simulation results after applying 1,000 random patterns is very close to that using GD. The Trojan cone in Fig. 4.3b consists of only AND gates, meaning that the probability of generating "1" at the output of Trojan gate T_{gj} is much less

Table 4.1 s298 and s344 benchmarks characteristics

Benchmark	# of inputs	# of flip-flops	The total number of inputs and flip-flops	The number of gates
s298	3	14	17	68
s344	9	15	24	71

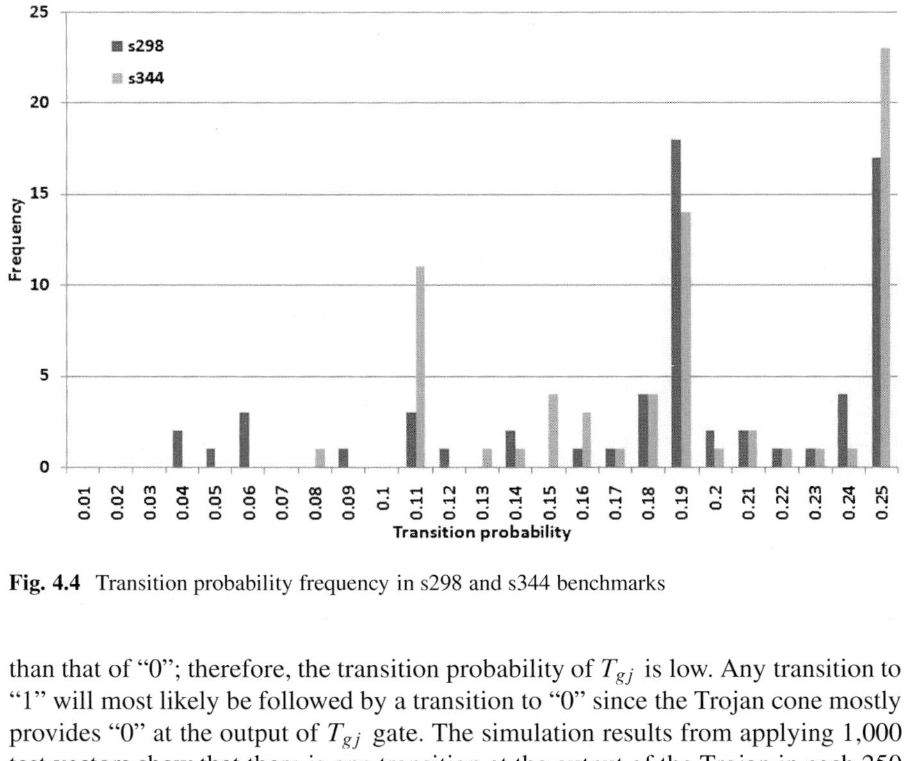

Fig. 4.4 Transition probability frequency in s298 and s344 benchmarks

than that of "0"; therefore, the transition probability of T_{gj} is low. Any transition to "1" will most likely be followed by a transition to "0" since the Trojan cone mostly provides "0" at the output of T_{gj} gate. The simulation results from applying 1,000 test vectors show that there is one transition at the output of the Trojan in each 250 clock cycles, and the probability analysis shows that every 255.6 clock cycles, one transition can be generated at the output of the T_{gj} gate.

Beside the interconnection among gates (i.e. circuit topology), transition probabilities of nets depend on the number of inputs and flip-flops. Primary inputs and flip-flops can determine a net's depth, which is the minimum distance of the net from either a primary input or a flip-flop. Such dependency is examined for two ISCAS'89 benchmarks (s298 and s344). Table 4.1 shows the benchmarks' characteristics. The benchmarks have roughly the same number of gates; however, their number of inputs and flip-flops are different. Primary inputs and flip-flops provide immediate access to internal parts of a circuit, and thus increase nets' transition probabilities. Figure 4.4 compares the transition probabilities frequency within these benchmarks.

Figure 4.4 shows that s344 benchmark, having more inputs and flip-flops, has more nets with high-transition probability. Further, simulation results of applying

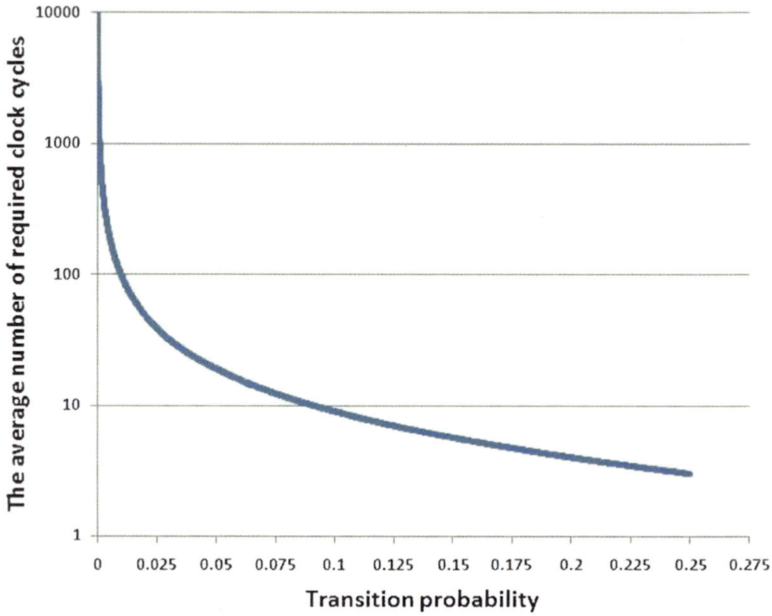

Fig. 4.5 The number of required clock cycles versus transition probability based on geometric distribution

random vectors in 1,000 clock cycles report 56,560 transitions in s344, and 44,600 in s298. Therefore, inserting dummy flip-flops can enhance accessibility to the internal parts of a circuit and can effectively increase the transition probability of circuit nets.

Both GD and simulation analyses show that as $P0$ or $P1$ of a net becomes too large or too small, the transition probability rapidly decreases. Therefore, to increase the transition probability, the $P0$ and $P1$ values should be close. The maximum net transition probability is $Pt = 0.25$ when $P0 = P1 = 1/2$. Given a cone's structure and the various types of gates in the cone, increasing transition probabilities seem impractical, but by adding dummy flip-flops to improve controllability, it is possible to increase transition probability for both $0 \rightarrow 1$ and $1 \rightarrow 0$ transitions. This is validated by the analysis performed using Geometric Distribution as shown in Fig. 4.5. As seen, the number of clock cycles to generate transitions increases exponentially as the transition probability decreases.

4.2 Dummy Scan Flip-Flop Insertion

When the probabilities of "1" and "0" of nets on a path within a cone become unidirectional (i.e. $P1 \gg P0$ or $P0 \gg P1$ similar to the example shown in Fig. 4.3b) the transition probability of the nets ($Pt_i = P_i0 \times P_i1$) rapidly decreases.

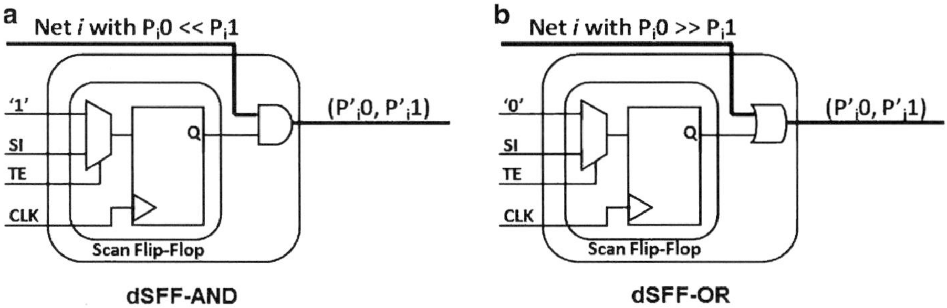

Fig. 4.6 The dummy flip-flop structures when (**a**) $P_i0 \ll P_i1$ and (**b**) $P_i0 \gg P_i1$

To ensure transition probabilities are greater than a specific threshold (P_{th}), dummy flip-flops can be inserted to bring the probabilities of "1" and "0" closer to each other. Note that both terms "dummy flip-flop" and "dummy scan flip-flop" refer to the increased controllability (transition probability) in a circuit.

Figure 4.6 shows the structure of a dummy scan flip-flop (dSFF) which consists of one scan flip-flop and one AND/OR gate. If the probability of "0" in the target net Net_i, P_i0, is less than the probability of "1" in Net_i, P_i1, then the output of scan flip-flop can be connected to an AND gate and Net_i can be restitched through the AND gate to increase P_i0, as depicted in Fig. 4.6a. However, if P_i1 is less than P_i0, an OR gate can be used to increase P_i1, as shown in Fig. 4.6b. In this work, *dSFF-AND* and *dSFF-OR* represent dummy scan flip-flops with AND and OR gates, respectively. Pairing a net with low transition probability with a dSFF increases the transition probability and nets at the fan-out. When the Test Enable (TE pin) is active, the output of a scan flip-flop is supplied by the Scan Input (the SI pin). The inserted dummy scan flip-flop has no impact on the functionality of the circuit. In the normal functional mode, the output of the scan flip-flop is supplied by either "0" for dSFF-OR or "1" for dSFF-AND to avoid changing the functionality of Net_i.

The probabilities of "1" and "0" at the output of a scan flip-flop are 1/2. Thus, by supplying internal nets with nets having equal "1" and "0" probabilities, the probabilities of "1" and "0" of target nets become closer, and their transition probabilities can be increased. Assume that P_i0 of Net_i is much greater than its P_i1, where

$$P_i0 = \frac{K}{N} \quad \text{and} \quad P_i1 = 1 - \frac{K}{N} \tag{4.1}$$

and K and N are cardinal values. K is the number of clock cycles in one experiment and N the number of desired outcomes. P_i0 approaches 1 (i.e. $K \approx N$) when it is assumed $P_i0 \gg P_i1$. By inserting one dSFF-OR, shown in Fig. 4.6b, the new probabilities are

$$P_i'0 = \frac{1}{2} \times P_i0 = \frac{K}{2N}, \text{ and } P_i'1 = 1 - \frac{K}{2N}. \tag{4.2}$$

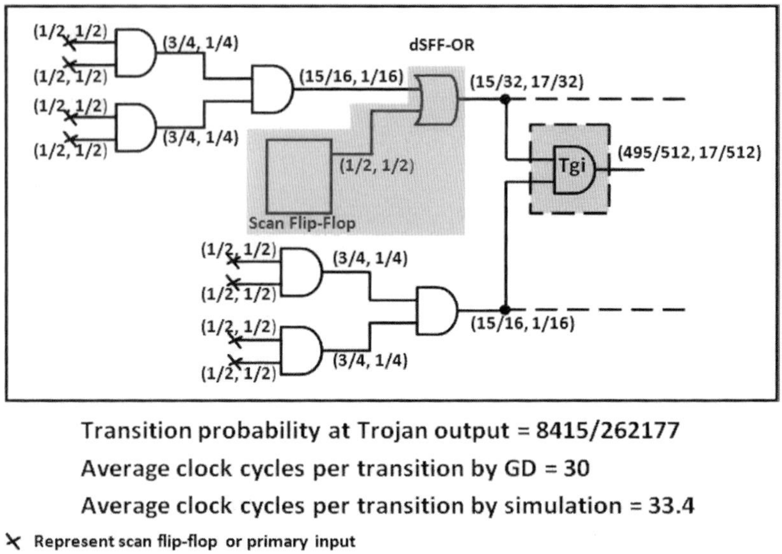

Fig. 4.7 Increasing transition probability by inserting dSFF-OR

As a result, $P_i'0$ will be smaller than P_i0 and $P_i'1$ will be greater than P_i1. Thus, after dummy flip-flop insertion, the transition probability of the target net will increase as

$$P_i'0 \times P_i'1 > P_i0 \times P_i1 \tag{4.3}$$

$$\frac{K}{2N} \times (1 - \frac{K}{2N}) > \frac{K}{N} \times (1 - \frac{K}{N}) \tag{4.4}$$

$$2N - K > 4(N - K) \tag{4.5}$$

which is true because $K \approx N$ and $N > K$.

Using the same analysis, it can be proven that by inserting a dSFF-AND when the P_i0 of a net is much lower than its P_i1, the transition probability can be increased.

Mathematical analysis shows that inserting a dSFF-OR on the upper-input net of T_{gi} gate in Fig. 4.3, as depicted in Fig. 4.7, reduces the number of clock cycles per transition from 255.6 to 30, on average, at the T_{gj} gate output. Furthermore, the simulation results closely confirm that there will be 33.4 clock cycles per transition, on average.

The TE pin is active during the test mode and a Trojan circuit can be designed to become active when the TE pin is inactive, which makes the dummy flip-flop technique ineffective. However, the *authentication* mode is different from the test mode, though it takes advantage of the testing infrastructure. In test mode, defects are targeted and different types of tests, such as the transition delay test, are used to detect them. But in authentication mode, a Trojan circuit is targeted and the detection objective is to reduce the Trojan partial/full activation time. For the purpose of authentication, it is not necessary always to keep the TE pin active. It can be switched between the authentication mode and the functional mode in each two successive clock cycles. During the authentication mode, patterns are shifted into scan flip-flops including dummy scan flip-flops, while during the functional mode the responses go into the scan flip-flops. The Trojan circuit is immediately exposed in one of two successive clock cycles in either the functional or authentication mode. The results for various switching patterns of the TE signal are shown in Sect. 4.4.

4.2.1 Removing Rare Triggering Conditions

Adversaries design Trojans to activate only under very rare conditions and circuit states; certain temperature or noise conditions, for example. Trojan design necessitates avoiding detection using structural or functional patterns. As an example, functional Trojans [7] can have $q \gg 1$ trigger inputs, which can (i) include nets with very low transition probabilities and/or (ii) create rare combinations. When the transition probability of Net_i is very low, P_i0 is much greater than P_i1 or vice versa, as discussed in Sect. 4.1. With q number of trigger inputs, the probability of generating a specific trigger vector is

$$P_{trigger-vector} = \prod_{i=1}^{q} P_i \tag{4.6}$$

where

$$P_i = \begin{cases} P_i0 \text{ for the trigger input } Net_i \text{ to be 0} \\ P_i1 \text{ for the trigger input } Net_i \text{ to be 1.} \end{cases} \tag{4.7}$$

$P_{trigger-vector}$ is expected to be very low if P_i0 or P_i1 is low. By inserting a dummy scan flip-flop, the nets' transition probability Increases, since P_i0 and P_i1 become closer. As a result, $P_{trigger-vector}$ also increases and the trigger vector will not be a rare event anymore. By increasing the transition probability of nets with low transition rates, hard-to-activate sites in a circuit will be eliminated. This will result in increasing the probability of switching within the Trojan circuit. If fully activated, the Trojan's output may change the main circuit functionality and it will be detected. In case of increasing switching in the Trojan, called *partial activation*, the Trojan can be detected much more easily using transient power or charge-based analysis techniques [8, 9, 12]. The dummy scan flip-flop insertion technique eliminates the need to focus on rare conditions, as proposed in [6, 13].

Table 4.2 The probability of Trojan activation before and after dSFF insertion in s38417 benchmark

	Before dSFF insertion			After dSFF insertion		
	$P0$	$P1$	$P_{Net1}0 \times P_{Net2}1$	$P0$	$P1$	$P_{Net1}0 \times P_{Net2}1$
Net 1	0.999995317077	4.6e−06	4.079e−06	0.989	0.011	0.094
Net 2	0.999959170737	4.08e−06		0.905	0.095	

For example, Table 4.2 shows the probability of two nets in s38417 benchmark before and after dummy scan flip-flop insertion. Assuming that a Trojan needs trigger vector {01} on $Net1$ and $Net2$, as seen in the table, the probability of the trigger vector is $P_{trigger-vector} = P_{Net1}0 \times P_{Net2}1 = 4.079e{-}06$ in the original circuit without dummy flip-flop. However, the probability increases to 0.094 after dummy flip-flop insertion.

4.2.2 Dummy Scan Flip-Flop Insertion Procedure

Figure 4.8 shows the dummy scan flip-flop insertion procedure. Nets with transition probabilities greater than a predefined transition probability threshold (P_{th}) and close to nets with transition probabilities lower than P_{th} are good candidates for dSFF insertion, since each of them can impact some other low transition nets at their fan-out cone.

After setting P_{th} and the main circuit as *CurrentCircuit* (Lines 1–2), the procedure will calculate the transition probability of all nets in the circuit (Line 3). Nets are then divided into two groups: (1) nets with transition probability higher than P_{th}, and (2) nets with transition probability lower than P_{th}. Nets in the first group obtained in Line 4 are then sorted and permanently stored in *SortedHighTransitionNets* (Line 6).

In the following, in Line 7 nets with transition probability less than P_{th} are identified and stored as *LowTransitionNets*. *NumberofLowTransitionNetsBefore*, in Line 8, saves the nets's number. The procedure, in Line 9, removes the net with the lowest/highest transition probability from *SortedHighTransitionNets*, depending upon *Order*. The removed net is restitched through dSFF in Line 11. Transition probability of nets after dSFF insertion is again calculated and the number of low transition nets is obtained. If the value is less than the number of low transition nets before dSFF insertion, the inserted dSFF is kept otherwise the dSFF would be ignored since no gain was obtained. In the following, if there is still any net with transition probability less than P_{th}, the procedure continues until there would not be neither any net with low transition probability nor any nets in *SortedHighTranstionNets*.

It is acknowledged that inserting dummy scan flip-flop increases the delay of paths and can impact circuit performance. Note that it is unlikely that

```
01: Set P_th (The desired threshold).
02: CurrentCircuit = SetDesign (the original circuit).
03: NetsProbability = NetsTransitionProbability (CurrentCircuit).
04: HighTransitionNets = Nets (1,Pth,NetsProbability).
05: Set Order "Increasing".
06: SortedHighTransitionNets = Sort (HighTransition,Order).
07: LowTransitionNets = Nets (P_th, 0, NetsProbability).
08: NumberofLowTransitionNetsBefore = Frequency (LowTransitionNets).
09: TargetNet = Pop (SortedHighTransitionNets).
10: dSFF = SelectDSFF (TargetNet).
11: InsertDSFF (dSFF, CurrentCircuit, TargetNet).
12: UpdatedCircuit = SetDesign (CurrentCircuit ).
13: NetsProbability = NetsTransitionProbability (UpdatedCircuit).
14: LowTransitionNets = Nets (P_th,0,NetsProbability).
15: NumberofLowTransitionNetsAfter = Frequency (LowTransitionNets).
16: If (NumberofLowTransitionNetsAfter < NumberofLowTransitionNetsBefore ) Then {
17:         CurrentCircuit = SetDesign (UpdatedCircuit).
18: }
19: If (NumberofLowTransitionNetsAfter > 0) Then {
20:         NumberofLowTransitionNetsBefore = NumberofLowTransitionNetsAfter;
21:         Go To Line 09;
22: } Else {Return CurrentCircuit.}
```

The description of functions :

Set Var1 Var2 : Sets the value of variable/constant Var2 to Var1.

SetDesign (Circuit) : Returns the Circuit variable.

NetsTransitionProbability (Circuit) : Obtains the Circuit variable and returns the transition probability of each net of the circuit.

Nets (High, Low, Nets) : Obtains an upper limit through a High variable, a lower limit through a Low variable, and a list of nets through Nets. Returns any net with a transition probability between the High variable and the Low variable.

Sort (Nets, Order) : Obtains a set of Nets and put them in order based on an Order variable which can have two values: Increasing, and Decreasing.

Frequency (Nets) : Returns the number of Nets.

Pop (Nets) : Removes the net at the top of Nets.

SelectDSFF (TargetNet) : Compare the probability of '1' and '0' for TargetNet. If the former probability is greater than the latter one, it returns dSFF-AND. Otherwise, it returns dSFF-OR.

InsertDSFF (dSFF, Circuit, Net) : Inserts a dSFF in the Cirucit and restitches Net through the dSFF.

Fig. 4.8 The dSFF insertion procedure

adversaries use nets on critical paths as input since it can both impact the path delay due to increased capacitance and can be easily detected using path delay fault test patterns. The procedure above allows one to avoid inserting dummy flip-flops on critical paths by eliminating nets on the critical paths from *HighTransitionNets*.

4.3 Transition Probability Threshold Analysis

Inserting dummy flip-flops to increase the transition probability of nets incurs area overhead. This area overhead is directed by the transition probability threshold (P_{th}). By setting P_{th}, the dummy scan flip-flop procedure ensures that all nets in the circuit have a transition probability greater or equal to P_{th}. P_{th} impacts both area overhead (i.e. the number of dSFFs) and the transition generation time in hardware Trojan gates. In general, setting a smaller P_{th} results in a smaller number of dSFFs but, on average, increases time to generate switching in Trojan gates. On the other hand, setting a larger P_{th} results in more dSFFs but reduces the transition generation time.

To set P_{th}, several parameters should be considered. They can be grouped into the two main categories of authentication and circuit parameters. Authentication parameters consist of two sub-parameters: (1) authentication time of each integrated circuit, T_{Au}, and (2) the clock period of tester, T_{Tester}. Circuit parameters consist of three sub-parameters: (1) the number of required transitions in the Trojan circuit, N_{Tr}, (2) the average number of clock cycles per transition which can be modeled using Geometric Distribution, and (3) circuit activity, $C_{activity}$. Note that N_{Tr} is an important parameter for power-based Trojan detection techniques since it indicates Trojan contribution into the total circuit power consumption. The larger the N_{Tr}, the easier the easier it is to detect a Trojan.

Equation (4.2) shows how authentication and circuit parameters are related:

$$T_{Au} \propto \frac{N_{Tr} \times T_{Tester} \times (P_{th}^{-1} - 1)}{C_{activity}} \quad (4.8)$$

T_{Au} is a user-defined parameter that depends on time-to-market and the criticality of the application in which the circuit will be used. The equation is based on the time-to-generate a specific number of transitions in a Trojan gate. With Geometric Distribution analysis, on average, $(P_{th}^{-1} - 1)$ clock cycles are required for each transition on nets whose transition probabilities are P_{th}. It is assumed that the inputs of Trojans are nets with transition probabilities of P_{th} in the equation to consider the worst authentication case. T_{Au} is inversely proportional to $C_{activity}$.

Equation (4.8) shows that T_{Au} directly depends on N_{Tr} and T_{Tester}:

1. Requiring more transitions in a Trojan circuit implies a longer authentication time.
2. The clock period of tester (T_{Tester}) determines how fast authentication patterns can be applied to a circuit under authentication (CUA). Applying patterns with higher frequency decreases T_{Au}.

P_{th} determines the transition probability threshold of the circuit; Geometric distribution causes P_{th} to increase and T_{Au} to decrease. Demonstrating a circuit with $C_{activity}$ in the "unit" and assuming $T_{Tester} = 4 \times 10^{-3}$ s, Fig. 4.9 shows that for a target authentication time, P_{th} and thus the area overhead, increases by adding to the number of required transitions in the Trojan circuit. Further, P_{th} decreases

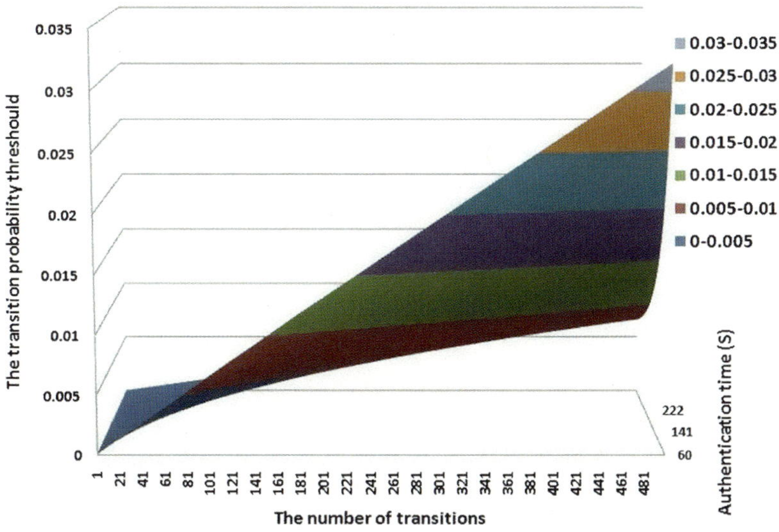

Fig. 4.9 Probability threshold versus authentication time and the number of transitions

by a specific number of transitions when the authentication time increases. The minimum P_{th} is obtained when the number of required transitions is the minimum and authentication time is the maximum.

Circuit activity ($C_{activity}$) is a function of a circuit's transition occurrence frequency, which is defined as

$$\sum_{Pt=0}^{0.25} (f_{Pt} \times Pt \times \sum_{Pt'=Pt}^{0.25} (f_{Pt'} \times Pt')) \tag{4.9}$$

where Pt and Pt' are transition probability, and f_{Pt} and $f_{Pt'}$ represent the number of nets with a transition probability of Pt and Pt' in the entire circuit. Equation (4.9) represents circuit characteristics by bringing the influence of nets with higher transition probabilities onto nets with lower transition probabilities. $C_{activity}$ is studied for s5378 and s641 benchmarks by inserting a NAND gate Trojan. Trojan inputs in the two benchmarks have roughly the same transition probabilities of ($P_t = 0.015$). Table 4.3 shows the benchmarks' characteristics and Fig. 4.10 presents their transition probability frequency. Although both circuits have the same number of inputs, s5378 is larger with a greater number of components consisting of gates and flip-flops. Transition occurrence frequencies of s5378 and s641 benchmarks are 578.84 and 124.92, respectively. It is expected that s5378 with higher $C_{activity}$ generates more transitions in the Trojan. Considering that $P_{th} = 10\text{e}{-}05$, simulation results show that there is one transition in every 29.8 clock cycles at the output of a Trojan in s5378, whereas there are 132.0 clock cycles in s641, on average.

T_{Au} depends on N_{Tr} and circuit characteristics in terms of $C_{activity}$ and P_{th}. N_{Tr} determines the number of required transitions to distinguish Trojan-inserted and

Table 4.3 s641 and s5378 benchmarks characteristics

Benchmark	# of inputs	# of flip-flops	# of components
s5378	35	176	823
s641	35	19	127

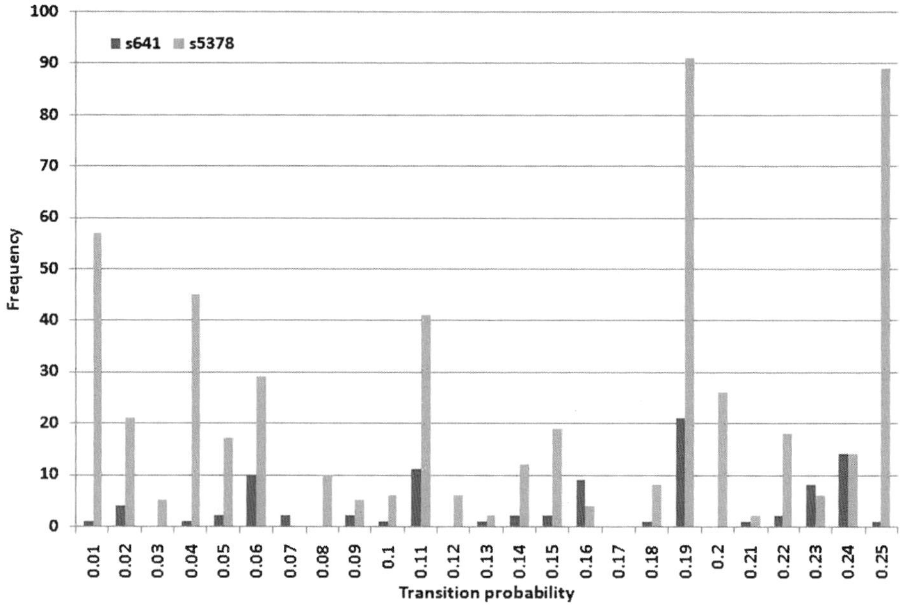

Fig. 4.10 Transition probability frequency in s641 and s5378 benchmarks

Trojan-free circuits. Circuits with higher $C_{activity}$ may increase Trojan activation and reduce T_{Au}. Furthermore, P_{th} provides an estimation of maximum T_{Au} by implying the rarest Trojan input vector application.

4.4 Simulation Results

Three programs carry out experiments. The first program calculates nets' transition probabilities, the second performs dummy flip-flop insertion, and the third enumerates the transitions after applying random patterns. The first program is written in TCL in TetraMAX [16]. It reads the circuit and calculates the probabilities of "1" and "0" in each net. Each net is either a primary input or the output of a gate. The probability of "1" and "0" for primary inputs are considered 1/2 (50 %) and for the output of gates are calculated based on the gates' functionality. As shown in Fig. 4.8, the second program is written in Perl [17] and performs dummy flip-flop insertion. The third program applies random patterns and monitors any transition on any net of the circuit using the Verilog Compiler (VCS) [16].

The dummy flip-flop insertion procedure is applied to the s38417 benchmark which contains 1564 flip-flops and 4933 gates. Four different transition probability thresholds are examined (P_{th}=10e−05, 10e−04, 10e−03, and 10e−2). The overhead, in terms of the number of dSFFs required to ensure all nets have transition probabilities higher than P_{th}, is evaluated. Assuming that P_{th}=10e−05, nets are divided into three groups: (1) Low Transition (LT) nets whose transition probabilities are less than 10e−05, (2) Medium Transition (MT) nets whose transition probabilities are between 10e−05 and 50e−05, and (3) High Transition (HT) probability nets whose transition probabilities are greater than 50e−05. Similar categorization is used for the other P_{th}s. To simulate the worst cases of Trojans activation, nets from the first and second categories are selected to be connected to the Trojans.

Four combinational comparator Trojan circuits, presented in Fig. 4.11, and one sequential Trojan, shown in Fig. 4.15, are inserted into the benchmark circuit. The Payload inputs, come from the Trigger (i.e. the trigger's output), and from the main circuit (i.e. data input). The comparator Trojans look for the rare combinations of inputs based on their "1" and "0" probabilities. Payload gates are selected based on the Trojan outputs' dominant values. Dash lines in the figure above represent the connection an adversary stitches between the main circuit and the Payload. Table 4.4 shows these Trojans' inputs. The first column shows the selected nets of the s38417 benchmark as Trojan inputs. The second and third columns indicate these nets' probabilities of "1" and '0,' and the last column represents their corresponding transition probabilities. The implemented Trojans are functional type and combinational [7]. They are activated conditionally and looking for rare trigger conditions. For example, Trojan 3, in Fig. 4.11, looks for (101011) whose probability of occurrence is about 0.4292e−20. The Trojans' outputs are passed to the main circuit and can cause functional failures.

The simulation results show the number of transitions over the entire main circuit, specifically transitions on the LT and MT nets. The total number of transitions at Trojan inputs and in Trojan circuits, and the number of transitions on Triggers' output that can potentially cause functional failure are reported. The effectiveness of dSFFs is measured with the Trojan-to-Circuit Activity (TCA) metric which is defined as the ratio of the number of transitions in the Trojan circuit (N_{Tr}) to the number of transitions in the entire design. Additionally, the number of transitions on the Payload output is also obtained and the difference between Payload output and its data input is investigated to further analyze the number of erroneous logic values injected into the main circuit.

When the value of the Trigger output is dormant (i.e. "1" for AND/NAND Payloads and "0" for OR/NOR Payloads), the Payload's output is the same as the Payload's data input; otherwise, the Payload's output depends on values of both the Trigger outputs and data inputs. If both are the same, then the output will be similar to both inputs. However, a different Payload input combination, assuming the Trojan is Triggered, means that the Payload output is due to Trigger input. This is called Trojan *full activation*, since the Payload output change (POC) can cause functional failure.

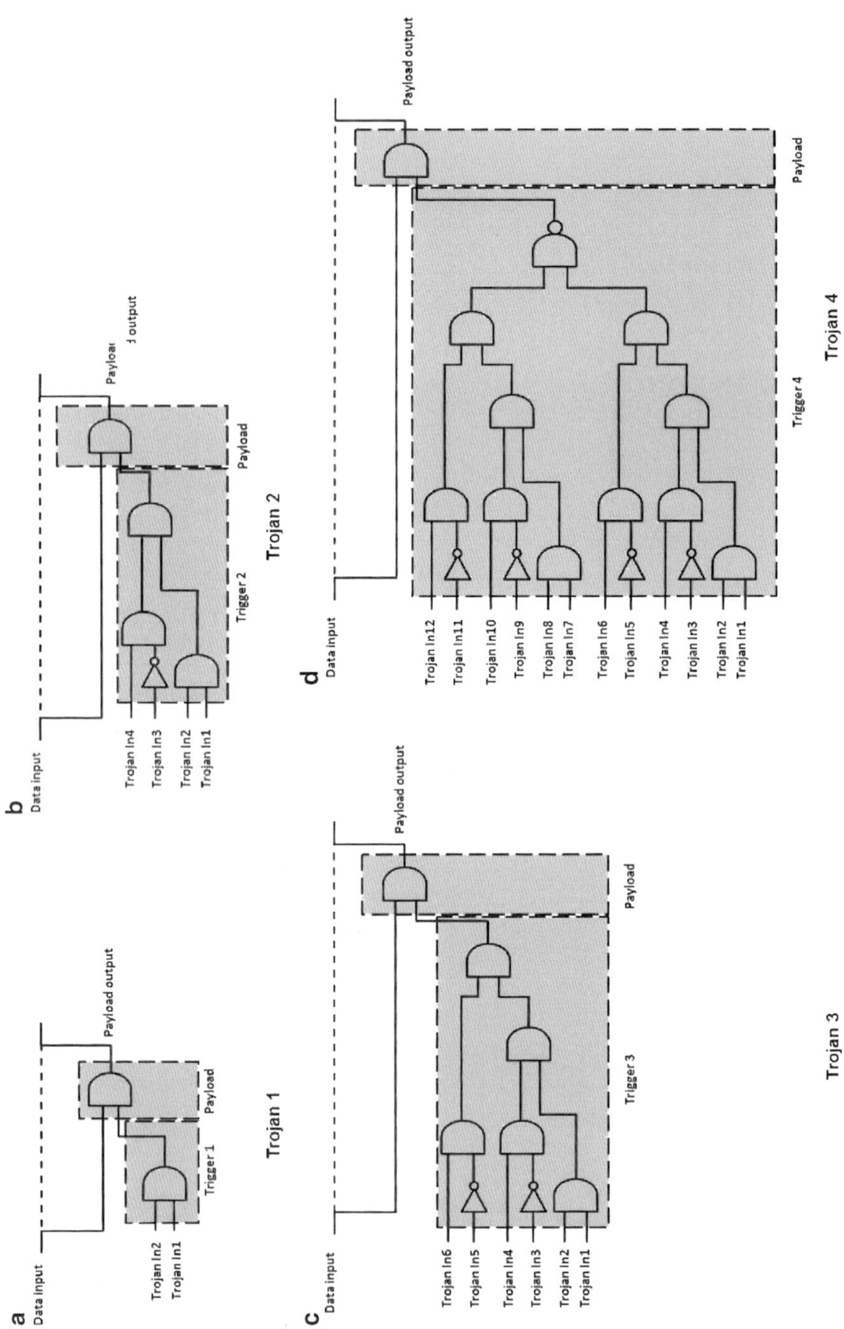

Fig. 4.11 Trojan circuits

Table 4.4 Trojan inputs characteristics

Net Name	$P0$	$P1$	Pt
Trojan In1 (n284)	0.9999	4.6829e−06	4.6829e−06
Trojan In2 (n382)	0.9999	4.6829e−06	4.6829e−06
Trojan In3 (n407)	4.5512e−05	0.9999	4.5510e−05
Trojan In4 (n578)	0.9999	4.6829e−06	4.6829e−06
Trojan In5 (n420)	0.9999	4.0829e−05	4.0827e−05
Trojan In6 (n309)	4.5512e−05	0.9999	4.5510e−05
Trojan In7 (n518)	0.9999	4.0829e−05	4.0827e−05
Trojan In8 (n616)	0.9999	4.0829e−05	4.0827e−05
Trojan In9 (n691)	0.9999	4.5512e−05	4.5510e−05
Trojan In10 (n766)	0.9999	4.5512e−05	4.5510e−05
Trojan In11 (n841)	0.9999	4.5512e−05	4.5510e−05
Trojan In12 (n505)	4.5512e−05	0.9999	4.5510e−05

The POC rate depends on the transition rate of Trigger output and Payload data input. When both Payload inputs have low transition probability the POC rate is unpredictable (small or large). For example, if Payload is an AND gate and data input and Trigger output have a high "1" probability, low POC rate is expected. On the other hand, if one of the Payload inputs has higher transition probability than the other, a larger POC rate is expected. Transition at the output of a gate based on the transitions of its inputs is analyzed in more detail. If Trigger output is active for many clock cycles, a large Payload output change is expected. The proposed method can help Trojan detection in two ways:

1. **Transient Power Analysis:** By increasing the number of transitions in a Trojan circuit, the technique can help improve the previously proposed power-based methods [5,6,9,11]. In this case, test vectors are applied in a test-per-clock (TPC) fashion since no observation is made by the flip-flops. Instead, the power pads and C4s are the observation points, since transient current is being measured. Suppose N_{sff} is the number of scan flip-flops and N_{vec_tpc} is the number of vectors; the total number of clock cycles $N_{total\_cycle} = N_{vec} + N_{sff} - 1$. When $N_{vec} \gg N_{sff}$, the total number of clock cycles equals the number of test vectors $N_{total\_cycle\_tpc} = N_{vec}$.
2. **Full Activation:** Increasing the probability of the full activation of a Trojan (making the data input different from Payload output) also increases the probability of observing an incorrect response to test vectors. In this case, the test vectors are applied in a test-per-scan fashion since the response of a test vector pair must be captured and scanned-out. The test vectors are applied similarly to launch-off-shift method used for delay testing except that there is no requirement on at-speed scan enable signal. The second vector is only a 1-bit shifted version of the first vector (i.e. initialization vector). If N_{sff} is the number of scan flip-flops and N_{vec_tps} is the number of vectors, the total number of clock cycles $N_{total\_cycle\_tps} = (N_{sff} + 1) \times N_{vec\_tps}$.

4.4.1 Without Dummy Flip-Flop

Simulations are run for $N_{vec\_tpc} = N_{vec\_tps} = 144$ test vectors. Table 4.5 shows transition statistics and the contribution of Trojans into the original circuit, i.e. before dSFF insertion. Column 2 shows the number of transitions in the entire circuit, including Trojans. In the next two columns, transition counts for LT and MT nets are reported. These numbers represent the net activities that are more likely to constitute Trojan inputs to make Trojan activation rare. The fifth column presents the number of transitions at Trojan inputs, implying attempts to activate Trojans subjected to various input combinations. Columns 6 and 7 show the number of transitions inside and at the output of Trojans, respectively. The total number of transitions in Trojans (N_{Tr}), the sum of transitions inside and at the output of Trojans, is reported in Column 8. Trojan contribution into the entire circuit is evaluated by a TCA metric and presented in Column 9. The last column (POC) indicates the number of Trojan full activations that result in functional errors inside the main circuit.

Table 4.5 shows that before dSFF insertion none of the Trojans is fully activated. The results indicate larger Trojans contribute more into the entire circuit activity (i.e. cause larger N_{Tr}), and thus have greater TCA, which is usually attributed to internal transitions.

4.4.2 P_{th} = 10e−05

There are four nets in s38417 benchmark with a transition probability less than 10e−05. 4 dSFFs are needed to increase these nets' transition probabilities beyond 10e−05. The 4 dSFFs make an area overhead of about 0.2 %. Table 4.6 shows increases in both circuit and Trojan activity. Although none of the Trojans is fully activated, there is an increase in the Trojans' TCA in proportion to their size. Furthermore, activity of LT and MT nets increases and is manifested in increasing Trojan activity. In the following, P_{th} is increased to 10e−04, and corresponding results are presented in Table 4.7.

4.4.3 P_{th} = 10e−04

The dSFF insertion procedure identifies 28 nets with a transition probability less than 10e−04. In this case, 16 dSFFs are inserted to ensure that these nets have a transition probability greater than P_{th}, incurring 0.8 % area overhead. The results in Table 4.7 show that LT and MT nets are more active by increasing P_{th}, compared with the previous cases. In addition, Trojans of smaller sizes are fully activated and cause functional errors in the main circuit several times, and larger Trojans

Table 4.5 Trojans activity analysis before dSFF insertion

Trojan	Total number of transitions	# of Transitions on LT nets	# of Transitions on MT nets	# of Transitions Trojan inputs	# of Transitions inside at Trojan circuit	# of Transitions at Trojan output	N_{Tr}	TCA	POC
Trojan 1	310,418	7	105	4	NA[a]	0	0	0.0E+00	0
Trojan 2	310,429	7	105	16	11	0	11	3.54E−05	0
Trojan 3	310,443	7	105	35	25	0	25	8.05E−05	0
Trojan 4	310,470	7	105	89	51	0	51	1.64E−04	0

[a]Not applicable

Table 4.6 Trojans activity analysis after dSFF insertion with $P_{th} = 10e{-}5$

Trojan	Total number of transitions	# of Transitions on LT nets	# of Transitions MT nets	# of Transitions on Trojan inputs	# of Transitions inside at Trojan circuit	# of Transitions at Trojan output	N_{Tr}	TCA	POC
Trojan 1	318,257	57	144	37	NA	1	1	3.13E−06	0
Trojan 2	318,284	57	144	64	14	0	14	4.39E−05	0
Trojan 3	318,299	57	144	97	28	0	28	8.79E−05	0
Trojan 4	318,365	57	144	163	95	0	95	2.98E−04	0

Table 4.7 Trojans activity analysis after dSFF insertion with $P_{th} = 10e{-}4$

Trojan	Total number of transitions	# of Transitions on LT nets	# of Transitions on MT nets	# of Transitions at Trojan inputs	# of Transitions inside Trojan circuit	# of Transitions at Trojan output	N_{Tr}	TCA	POC
Trojan 1	327,293	149	399	69	NA	18	18	5.49E−05	14
Trojan 2	327,322	149	399	148	73	7	80	2.44E−04	9
Trojan 3	327,357	149	399	190	116	0	116	3.54E−04	6
Trojan 4	327,436	149	399	361	195	0	195	5.95E−04	0

bring forth more internal transitions. Consequently, there is high activity in Trojan circuits and a significant increase in the Trojans' TCA. To verify that continuously increasing P_{th} increases a Trojan's contribution, P_{th} is increased to 10e–03.

4.4.4 P_{th} = 10e–03

Increasing transition probability of nets beyond 10e–03 requires 60 dSFFs and imposes 3.0 % overhead. The results in Table 4.8, contrary to what was expected, show decreases in both main circuit and Trojans activity. LT and MT nets are less active, compared with P_{th} = 10e–04, and as a result, Trojans become less activated. Even the total number of transitions in the entire circuit decreases. The exception is the TCA factor of Trojan 4 which increased, although it becomes less active due to greater decreases in the total number of transitions within the entire circuit. Detail analysis in the following shows that increasing the nets' transition probability beyond a specific threshold does not necessarily increase the number of transitions in the entire circuit.

Any circuit consists of primary gates, mainly NAND and NOR gates, and any other complex gate and module can be made using these primary gates. A transition at the output of a gate is a function of a transition on its inputs. Figure 4.12 shows transition probability at the output of 2-input NAND and 2-input NOR gates based on the transition probability of their inputs. The maximum transition probability of a net is 0.25 and is obtained when net probabilities of "1" and "0" are equal to 1/2. However, Fig. 4.12 indicates that maximum transition probability at the output of the gates is when the transition probability of one of its inputs is high and that of the other input is low. This trend can be seen in both NAND and NOR gates and is observed for both AND and OR gates. Further, Fig. 4.12 indicates when transition probabilities of inputs are both 0.25 (the maximum value), the transition probabilities at gates' outputs are 0.1875 in the both gates. In sum, increasing P_{th} to increase the transition probability of individual nets may not necessarily increase the number of transitions in the entire circuit. To confirm this fact P_{th} is increased to 10e–02 and results are presented in Table 4.9.

4.4.5 P_{th} = 10e–02

Incurring 5.2 % area overhead, 100 dSFFs are needed to create a transition probability of all nets above 10e–02. As expected, the total number of transitions in the entire design decreases more and the number of transitions inside and at the output of Trojan circuits is less than the results for P_{th} = 10e–04. However, since the number of transitions in the entire circuit is roughly half of corresponding values with P_{th} = 10e–04, there is an increase in the Trojans' TCA with P_{th} = 10e–02.

Table 4.10 presents transition statistics in the entire circuit and the LT and MT nets at examined P_{th}s. The results in Column 2 indicate increasing P_{th} decreases the

Table 4.8 Trojans activity analysis after dSFF insertion with $P_{th} = 10e{-}3$

Trojan	Total number of transitions	# of Transitions on LT nets	# of Transitions on MT nets	# of Transitions at Trojan inputs	# of Transitions inside Trojan circuit	# of Transitions at Trojan output	N_{Tr}	TCA	POC
Trojan 1	235,378	46	243	30	NA	11	11	4.67E−05	3
Trojan 2	235,394	46	243	42	38	1	39	1.65E−04	0
Trojan 3	235,418	46	243	96	62	0	62	2.63E−04	0
Trojan 4	235,523	46	243	186	167	0	167	7.09E−04	0

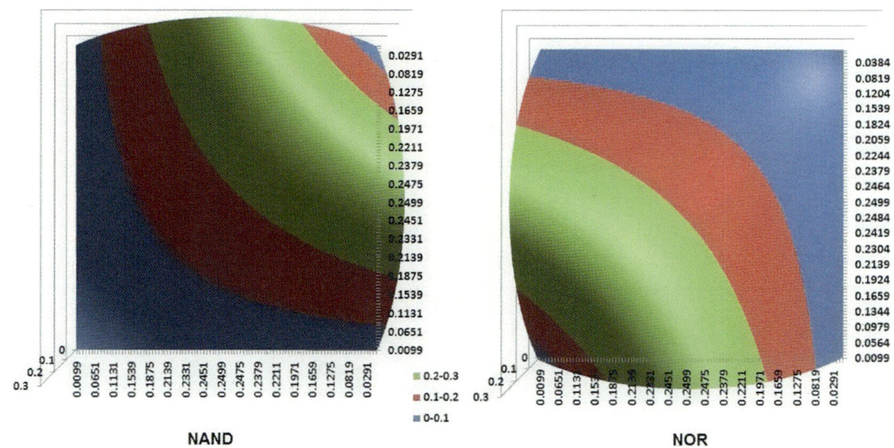

Fig. 4.12 Transition density analysis for NAND and NOR gates

Table 4.9 Trojans activity analysis after dSFF insertion with $P_{th} = 10e{-}2$

Trojan	Total number of transitions	# of Transitions on LT nets	# of Transitions on MT nets	# of Transitions at Trojan inputs	# of Transitions inside Trojan circuit	# of Transitions at Trojan output	N_{Tr}	TCA	POC
Trojan 1	185,875	139	383	67	NA	10	10	5.38E−05	5
Trojan 2	185,915	139	383	132	67	2	69	3.71E−04	0
Trojan 3	185,946	139	383	176	100	0	100	5.37E−04	0
Trojan 4	186,049	139	383	352	204	0	204	1.09E−03	0

Table 4.10 The transitions statistics in the entire design and MT and LT nets

Pth	The number of transitions in the entire design	The transitions percentage of MT and LT nets	Average number of transitions on LT and MT nets	Average number of transitions per clock in the entire design	Average number of transitions on each net
Before dSFF insertion	310,418	3.6e−02 %	5.6	2,155	66
10e−05	318,527	6.3e−02 %	10.0	2,212	68
10e−04	328,684	1.5e−01 %	25.4	2,283	70
10e−03	236,314	1.2e−01 %	14.4	1,641	50
10e−02	185,875	2.8e−01 %	26.1	1,290	39

Table 4.11 P_{th} analysis

Pth	10e−05	10e−04	10e−03	10e−02
The number of nets	4	28	129	275
The number of dSFFs	4	16	60	100
Area overhead (%)	0.2	0.8	3.0	5.2
The ratio of # dSFF to # nets (%)	100	57	46	36

number of transitions in the entire design, caused by transition characteristics of the primary gates. On the other hand, Column 3 shows an increase in the percentage of transitions on LT and MT nets. In other words, there is a transition movement from HT nets to MT and LT nets. The next column also demonstrates that by increasing P_{th}, the average number of transitions on the LT and MT nets will increase. The last two columns corroborate that by increasing P_{th}, there is decrease in the number of transitions in the entire design per clock and on each net on average.

In Table 4.11, the results show that by increasing P_{th}, although there are more low transition nets, the number of required dSFFs is decreased. Further, the simulation results show that smaller Trojans, e.g. Trojan 1 and Trojan 2, can be fully activated more frequently. while they offer a smaller contribution to circuit activity. On the other hand, larger Trojans, e.g. Trojan 3 and Trojan 4, are harder to fully activate and contribute more to circuit activity.

4.4.6 TE Attack Analysis

An adversary may design a Trojan to be inactive during authentication time when the TE signal is active. The Trojan may use the TE signal as a trigger input and start operating when the TE signal is inactive, i.e. when the circuit is in functional mode.

As a countermeasure, the TE signal must frequently be switched on and off. Figure 4.13 presents this basic idea with three alternating scenarios: (1) TE1(N_{vec_tpc})0(0), (2) TE1(1)0(1), and (3) TE1(m)0(1). In the first scenario, TE signal is on (high) and the circuit is in shift (or scan) mode during the entire authentication time. TE1(1)0(1), in the second scenario, represents the case where TE=1 for one clock cycle (a random bit is shifted into the scan chain) and TE=0 in the next clock cycle (the response goes into scan chain). In the TE1(m)0(1) scenario, the TE signal is on for m clock cycles and then gets off for one clock cycle.

To evaluate the effectiveness of alternating the TE signal, Trojan 3 is equipped with the TE signal such that it is functional only when TE signal is off. P_{th} is set to 10e−4 and five cases are simulated:

1. TE1(1)0(1): the TE signal is switched at each clock cycle.
2. TE1(10)0(1): the TE signal is on for $m = 10$ clock cycles and then switches off for one clock cycle.

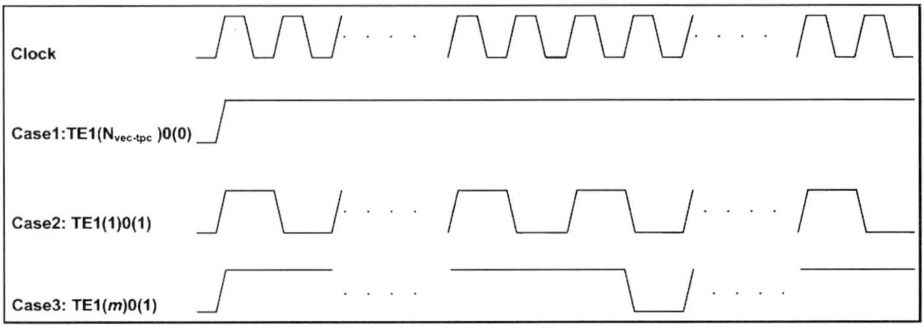

Fig. 4.13 Different application of the test enable signal

3. TE1(20)0(1): on state of the TE signal lasts for $m = 20$ clock cycles and then the TE is switched off for one clock cycle.
4. TE1(30)0(1): for $m = 30$ clock cycles the TE signal is on and then is switched off for one clock cycle.
5. TE1(144)0(0): the TE signal is kept high for the entire simulation, $m = 144$.

These simulations are run three times, and Table 4.12 shows the average results. The results show that the total number of transitions in the circuit increases with increasing m. Accordingly, the LT and MT net activity increases, and it augments the number of transitions on the Trojan's inputs. However, the results also show that the number of transitions inside the Trojan consistently decreases by increasing m, and the Trojan is never fully activated (POC=0). As a result, the TCA for the TE1(1)0(1) case is the largest and decreases by increasing m. Moreover, Table 4.12 shows that switching the TE signal with each clock cycle provides comparable TCA with the case of Trojan 3 in Table 4.7. Therefore, Trojan impact is exposed by switching the TE signal on and off, even when Trojan is designed such that it only operates when the TE signal is inactive.

4.4.7 Transient Power Analysis

The effectiveness of dummy flip-flops in power-based techniques is studied by analyzing the contribution of Trojan 4 to circuit power consumption. Two designs are generated: (1) design without a dummy flip-flop and (2) design with $P_{th} = 10e{-}4$. These designs and their corresponding Trojan-free ones are implemented in Synopsys' Astro and then their Spice netlists are extracted using Synopsys' StarRCXT [16]. To analyze the contribution of the Trojan on a circuit's power consumption, Trojan-to-Circuit Charge consumption (TCC) is measured per positive level of clock cycle. TCC is defined as

Table 4.12 Alternating test enable signal analysis

Trojan	Total number of transitions	# of Transitions on LT nets	# of Transitions on MT nets	# of Transitions at Trojan inputs	# of Transitions inside Trojan circuit	# of Transitions at Trojan output	N_{Tr}	TCA	POC
TE1(1)0(1)	25,689	31	364	102	17	0	17	6.97E−04	0
TE1(10)0(1)	40,003	59	372	139	4	0	4	1.07E−04	0
TE1(20)0(1)	42,220	113	498	209	2	0	2	6.17E−05	0
TE1(30)0(1)	43,617	154	567	258	1	0	1	3.17E−05	0
TE1(144)0(0)	47,246	206	659	317	0	0	0	0.00E+00	0

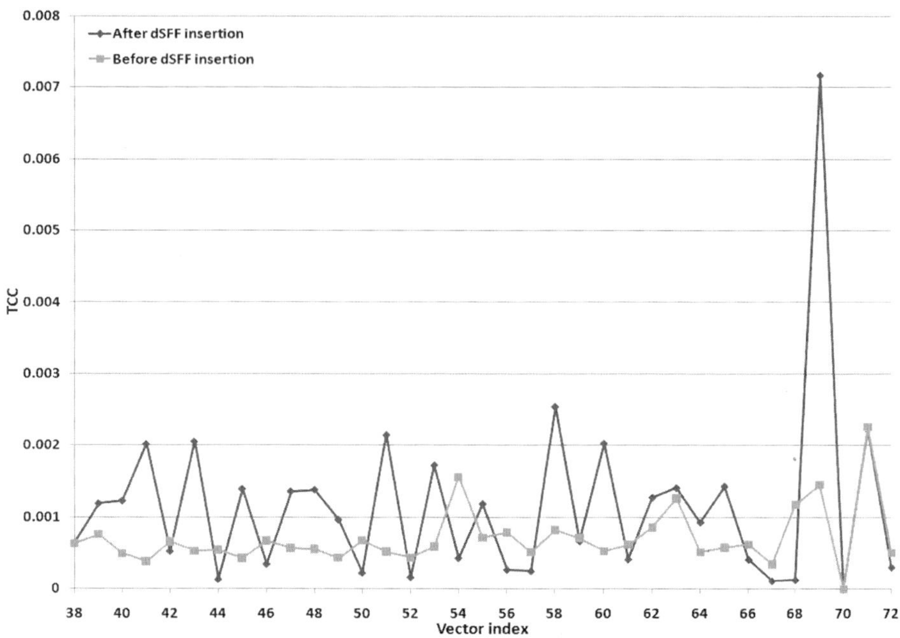

Fig. 4.14 The TCC of Trojan 4 before dSFF insertion and after it with $P_{th} = 10\text{e}{-04}$

$$TCC = \frac{\int_0^{T/2} I_{Trojan}(t)\, dt}{\int_0^{T/2} I_{Circuit}(t)\, dt} \tag{4.10}$$

where T is the clock period, $I_{Trojan}(t)$ denotes Trojan current consumption (which is the difference between Trojan-inserted and Trojan-free circuits), and $I_{Circuit}(t)$ denotes Trojan-inserted circuit current consumption.

Figure 4.14 shows TCC before and after dSFF insertion for vectors 38–72. The results show that the Trojan impact is magnified after dSFF insertion in most cases when compared with the circuit without dSFF. The results indicate that per clock Trojan contribution when dSFF is used is on average 2 times more when compared with the case without dSFF. Moreover, there are a number of cases where TCC after dSFF insertion is significantly greater than before dSFF insertion, and it helps detect the Trojan even in the presence of process variations. The impact of the Trojan can be further magnified using the charge integration method proposed in [12]. Moreover, it can be concluded that TCA, calculated at the logic level, is a true representation of TCC, when measured at circuit level. As when there is an increase in TCC measurement for Trojan 4 from before to after dSFF insertion, TCA of Trojan 4 increases from before dSFF insertion, as indicated in Table 4.5, to after dSFF insertion, shown in Table 4.7.

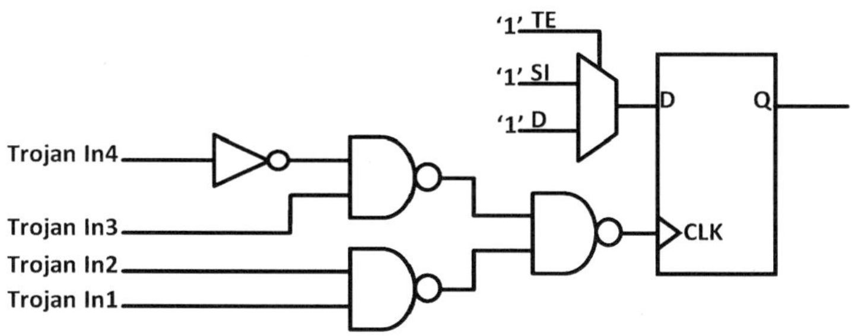

Fig. 4.15 A sequential Trojan circuit

4.4.8 Sequential Trojan Analysis

Trojans can be sequential and use memory elements, such as flip-flops or latches, to implement a finite state machine. It is expected that sequential Trojans have considerable impact on circuit power consumption. A memory element consists of several gates, such as AND and INV, which can incur extra capacitance load on the clock tree.

An adversary can eliminate Trojan impact on the clock tree by supplying Trojan clock input through a Trojan cone. Figure 4.15 presents a sequential Trojan without payload which consists of one scan flip-flop whose TE, scan-in (SI), and data (D) input signals are all stuck at "1" to merely analyze clock input (CLK) contribution. The CLK input is supplied by a Trojan cone as depicted in the figure, and the Trojan cone inputs are the same as inputs of Trojan 2 in Fig. 4.11.

The sequential Trojan is inserted before dSFF insertion and after dSFF insertion with $P_{th} = 10e{-}04$. Figure 4.16 show TCC measurements for vectors 38–68. The results show that dSFF insertion can significantly increase Trojan contribution to the circuit power consumption. Comparing the TCC of a sequential Trojan in Fig. 4.16 and that of Trojan 4 in Fig. 4.14 indicates that sequential Trojans have greater impact on circuit power consumption compared to the combinational Trojans, even though sequential Trojans may include fewer gates.

4.5 Summary

The topology of a circuit and the number of primary inputs and flip-flops determine the switching activity of the circuit. Transitions can be modeled using GD, and the average number of clock cycles required to generate a transition is estimated. Furthermore, it is shown that inserting dummy scan flip-flop can reduce the time needed to generate a transition. This realization develops a dummy flip-flop insertion procedure aimed at augmenting net transition probabilities in a circuit, and increasing the activity of hardware Trojans in Integrated Circuits. The simulation

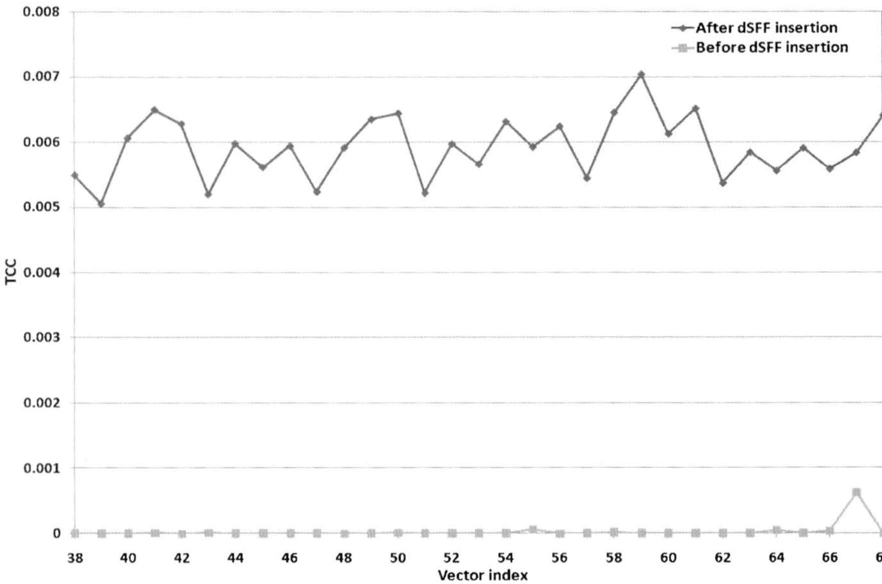

Fig. 4.16 The TCC of sequential Trojan before dSFF insertion and after it with $P_{th} = 10\text{e}{-}04$

results for s38417 benchmark demonstrate that it is possible to significantly increase switching activity in Trojan circuits. Smaller Trojans may be fully activated and cause functional failures, and larger Trojans contribute more to side-channel signals and are detected as an abnormality.

References

1. Y. Jin and Y. Makris, "Hardware Trojan detection using path delay fingerprint," in Proc. of the *IEEE International Workshop on Hardware-Oriented Security and Trust (HOST 2008)*, pp. 51–57, 2008.
2. J. Li and J. Lach, "At-speed delay characterization for IC authentication and Trojan Horse detection," in Proc. of the *IEEE International Workshop on Hardware-Oriented Security and Trust (HOST 2008)*, pp. 8–14, 2008.
3. M. Tehranipoor and F. Koushanfar, "A Survey of Hardware Trojan Taxonomy and Detection," in *IEEE Design and Test of Computers*, pp. 10–25, 2010.
4. M. Banga, M. Chandrasekar, L. Fang, and M. S. Hsiao, "Guided test generation for isolation and detection of embedded trojans in ICs," in Proc. of the *in Proceedings of the IEEE/ACM Great Lakes Symposium on VLSI*, pp. 363–366, April 2008.
5. M. Banga and M. S. Hsiao, "A region based approach for the identification of hardware trojans," in Proc. of the *IEEE International Workshop on Hardware-Oriented Security and Trust (HOST08)*, pp. 40–47, June 2008.
6. R. S. Chakraborty, F. Wolff, S. Paul, C. Papachristou and S. Bhunia, "MERO: A Statistical Approach for Hardware Trojan Detection," in Proc. of the *Workshop on Cryptographic Hardware and Embedded Systems (CHES)*, 2009.

7. X. Wang, M. Tehranipoor and J. Plusquellic, "Detecting malicious inclusions in secure hardware: Challenges and solutions," in Proc. of the *IEEE International Workshop on Hardware-Oriented Security and Trust (HOST 2008)*, pp. 15–19, 2008.
8. D. Agrawal, S. Baktir, D. Karakoyunlu, P. Rohatgi and B. Sunar, "Trojan Detection using IC Fingerprinting," in Proc. of the *Symposium on Security and Privacy*, pp. 296–310, 2007.
9. R. Rad, X. Wang, J. Plusquellic and M. Tehranipoor, "Power Supply Signal Calibration Techniques for Improving Detection Resolution to Hardware Trojans," in Proc. of the *International Conference on Computer-Aided Design (ICCAD08)*, pp. 632–639, 2008.
10. D. D. Wackerly, W. Mendenhall III and R. L. Scheaffer, "Mathematical Statistics with Application, 7th edition" Thomson Learning, Inc., 2008.
11. L. Wang, C. Stroud, N. Touba, " System-on-Chip Test Architecture: Nanometer Design for Tetsability" Morgan Kaufmann Publishers.
12. X. Wang, H. Salmani, M. Tehranipoor and J. Plusquellic, "Hardware Trojan Detection and Isolation Using Current Integration and Localized Current Analysis," in Proc. of the *International Symposium on Fault and Defect Tolerance in VLSI Systems (DFT08)*, pp. 87–95, 2008.
13. F. Wolff, C. Papachristou, S. Bhunia and R.S. Chakraborty, "Towards Trojan-Free Trusted ICs: Problem Analysis and Detection Scheme," in Proc. of the *Design, Automation and Test in Europe(DATE '08)*, pp. 1362–1365, 2008.
14. R. Sankaralingam, R. R. Oruganti and N. A. Touba, "Static Compaction Techniques to Control Scan Vector Power Dissipation," in Proc. of the *IEEE VLSI Test Symposium (VTS'00)*, pp. 35–40, 2000.
15. H. Salmani, M. Tehranipoor, and J. Plusquellic, "New Design Strategy for Improving Hardware Trojan Detection and Reducing Trojan Activation Time," in Proc. of the *IEEE symposium on Hardware-Oriented Security and Trust (HOST 2009)*, pp. 66–73, 2009.
16. Synopsys Inc., User Manual, http://www.synopsys.com/.
17. L. Wall, T. Christiansen, and J. Orwant, *Programming Perl (3rd Edition),* Oreilly & Associates Inc., 2000.

Chapter 5
Design for Hardware Trust: Layout-Aware Scan Cell Reordering

Side-channel signal analysis techniques based on transient power have proven highly effective in extracting information about the internal operations of a circuit [1,2]. In power-based side-channel signal analysis, it is possible to extract a Trojan signal by monitoring power pads/ports, even in the presence of various types of noise including measurement noise, ambient noise, and other random signal variations that manifest during the circuit operation [5].

To avoid easy detection, an adversary may design a Trojan to have little impact on circuit power. Developing a pattern generation strategy to localize switching activity and reduce background noise (i.e. circuit power) is an extremely challenging task. Using regional activation to limit transitions in a target region of a circuit while keeping the rest of the circuit quiet is an effective way of increasing the Trojan detection resolution in terms of Trojan-to-circuit switching activity and Trojan-to-circuit power consumption. Trojan-to-circuit switching activity (TCA) is defined as the ratio of the number of transitions inside the Trojan circuit to the number of transitions over the entire circuit. Trojan-to-circuit power consumption (TCP) is defined as the ratio of Trojan power consumption to circuit power consumption.

This chapter introduces a hardware trust architecture to magnify functional Trojan activity. As there is a high correlation between the number of transitions in circuit scan cells and switching activity within the circuit, the proposed architecture reorders scan cells based on their placement during physical design to reduce switching activity in the main circuit by limiting it to a specific region. This helps magnify Trojan contribution to the total circuit transient power by increasing TCA and TCP. The proposed technique aims to improve the efficiency of power-based side-channel signal analysis techniques in detecting hardware Trojans.

5.1 Scan Cell Reordering

In general, transitions in a circuit are mainly caused by transitions at primary inputs and scan flip-flops. In a large circuit, the capability of primary inputs to generate transition is restricted to first levels of the circuit. However, the scan architecture

M. Tehranipoor et al., *Integrated Circuit Authentication: Hardware Trojans and Counterfeit Detection*, DOI 10.1007/978-3-319-00816-5_5,

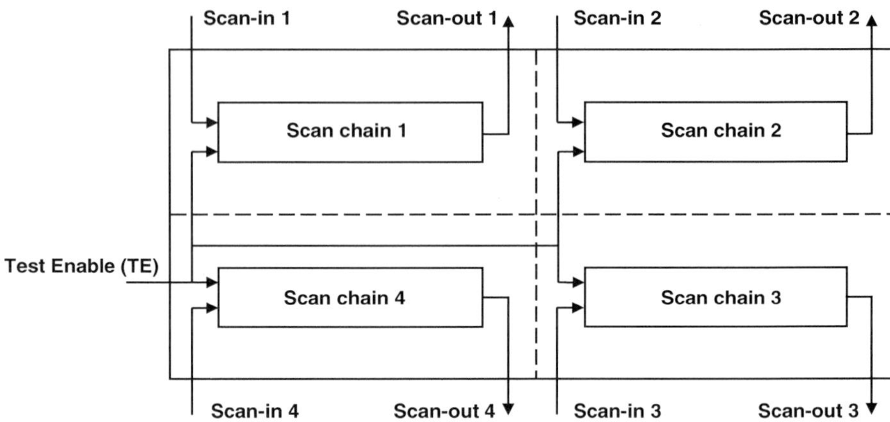

Fig. 5.1 The layout-aware scan-cell reordering concept

allows easy access to the internal parts of the circuit. Furthermore, there is a high correlation between the total power consumption of a circuit under test (CUT) and the total number of transitions in scan cells during scan-based testing [8, 9].

Scan-cell reordering techniques have already been proposed to reduce power during the scan test [10, 11], to enhance delay fault coverage [12, 13], or to minimize scan paths [14, 15]. In [10], using a set of scan cells and a given sequence of deterministic test vectors, a heuristic provided a scan chain order that minimizes the occurrence of transitions and hence the peak power during testing. A scan-cell reordering scheme has been proposed in [11] to reduce net transitions during the shifting-in and shifting-out of a given test pattern set. Authors in [12] have presented a restricted scan chain reordering technique to enhance delay fault coverage. The proposed technique restricts the distance by which a scan flip-flop can be moved to create the new scan chain order. In [14], an approach has been proposed to reorder scan chains, based on physical design information, to reduce routing bottleneck and minimize design constraint violations. The proposed method reorders scan cells within scan chains so that the total inter-clusters Manhattan distance is minimized. In this chapter a new scan cell reordering method is proposed for improving Trojan detection based on power-based signal analysis. In general, scan cells are scattered across the circuit layout, and many gates are activated at the same time across the layout during IC authentication. Reordering of scan cells based on their geometric positions can significantly restrict switching activity to a specific region.

Figure 5.1 shows the basic concept of the proposed layout-aware scan-cell reordering. Assume that a design with four scan chains ($N = 4$) is divided into four regions. The method forms the scan chains such that the scan cells placed in each selected region are connected to each other. This is to ensure that the scan chains have the same length, but that is not a requirement. The technique enables the magnification of Trojan impact by increasing the Trojan-to-circuit power

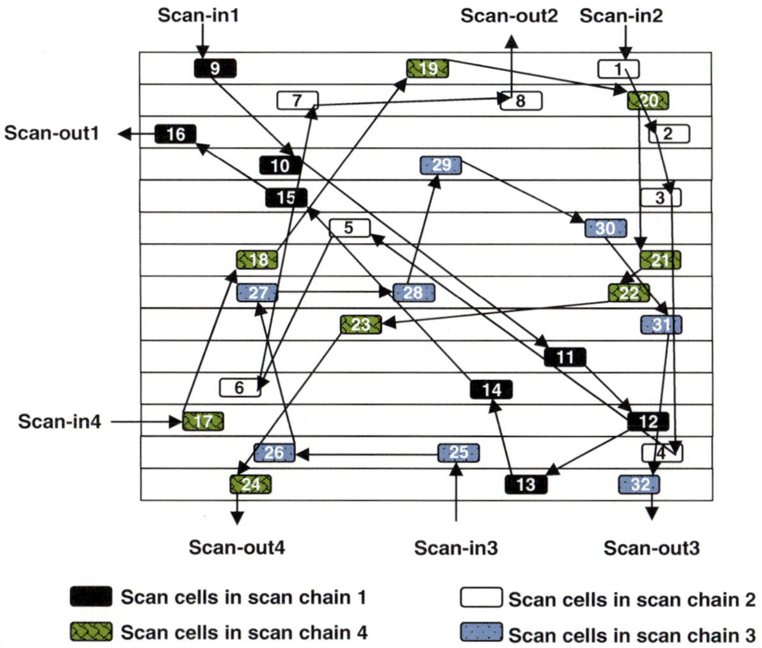

Fig. 5.2 The traditional scan chain organization in s838 benchmark

consumption ratio by maximizing switching in the target region (e.g. the region containing scan chain 4) while minimizing switching in all the other regions (1, 2, and 3).

As an example, Fig. 5.2 shows the organization of scan chains in the small ISCAS'89 s838 benchmark, where 32 scan cells are grouped into four scan chains using the Synopsys' Design Compiler [16]. The figure shows that scan chains are scattered across the layout, and the entire design is subject to dispersed transitions by using any scan chain. The proposed procedure groups scan cells by their placement in the layout. Due to the lack of placement information at the front-end phase during scan chain insertion, it is not possible to group scan cells and arrange them based on geometric information. Therefore, the layout-aware scan-cell reordering is performed after placement and before routing.

The reordering procedure obtains scan cells' placement information and re-stitches them. Although the basic idea is applicable in any design environment, here, the procedure is implemented by using the Synopsys' Astro [16]. First, the placement information of scan cells is extracted. Then, existing connections between scan cells are removed. In the following, a circuit is divided into N regions. Divided into one or more scan chains inside a region, scan cells are connected together while connections are optimized to reduce routing congestion. Finally, the circuit netlist is updated with re-stitched cells to be considered for routing.

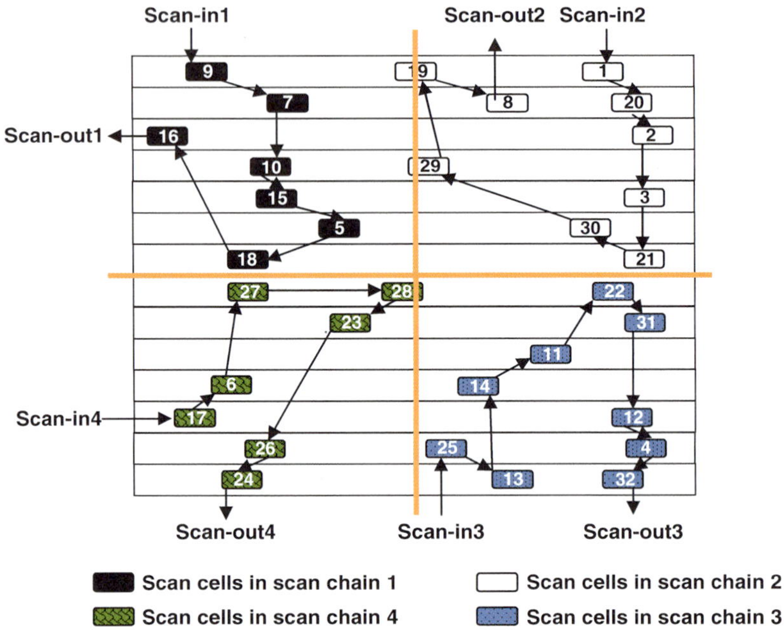

Fig. 5.3 The scan chain organization in s838 benchmark using the layout-aware scan-cell reordering

Regions are controlled by scan-chains; therefore, the number of scan chains may determine N. Given the limitation on the number of pins used for testing, when the number of regions is large, a compression-like architecture including a phase shifter can be used. The technique has no impact on the pattern generation flow and fault coverage, and it does not cause area and pin overhead.

Figure 5.3 shows the new organization of scan chains in s838 benchmark after performing the layout-aware scan-cell reordering. In this example, the scan cells are grouped into four regions, N=4, as if the circuit layout is divided by 4 based on the cell locations. The effectiveness of the scan-cell reordering technique in limiting switching activity in any target region is evaluated for larger ISCAS'89 benchmarks, s38417 with 1564 flip-flops and 4933 gates, and s35932 with 1728 flip-flops and 3926 gates. Using the layout-aware scan-cell reordering method, scan cells in both benchmarks are grouped into $N = 4$ regions with each benchmark consisting of 48 scan chains; that is, each region consists of 12 scan chains. The simulation is run four times, and each run consists of three pattern sets. Each time, different random patterns are applied. Each pattern set consists of 41 test vectors in s38417 and 46 test vectors in s35932. Patterns apply random "0"s and "1"s to scan chains covering the target region at the bottom left corner of the benchmarks' layouts, while a "0" is applied to all other scan chains. To increase randomness, the circuit is always set to the scan mode by keeping the test-enable (TE) signal active. As there is a

Region 1	Region 2
s38417: 15.7%, 15.8%, 15.2%, 15%	s38417: 7.1%, 7.2%, 7.2%, 7%
s35932: 11.6%, 11.4%, 11.9%, 11%	s35932: 8.3%, 7.5%, 8.2%, 6.8%
Region 4	**Region 3**
s38417: 67.5%, 67.7%, 69%, 68.3%	s38417: 9.7%, 9.3%, 9.6%, 9.5%
s35932: 69.7%, 71%, 69.3%, 73%	s35932: 10.3%, 9.9%, 10.4%, 9%

Fig. 5.4 The percentage of switching activity in each region of s38417 and s35932 after running four simulations. The results are shown as (Run1, Run2, Run3, Run4) for each benchmark

high correlation between circuit switching activity and scan flip-flops transitions, the circuit is subjected to high randomness by controlling the values of scan flip-flops through scan chains. Other cases, such as using TE signal and the functional capture clock, can also be applied a in case where the adversary uses the TE signal to deactivate a Trojan in the authentication mode. The percentage of activity in each region of benchmarks is reported in Fig. 5.4 as (Run1, Run2, Run3, Run4). The results clearly indicate that in all four runs, switching activities are mostly limited to the target region labeled "Region 4" in Fig. 5.4, while the other regions are kept fairly inactive in both benchmarks. Note that the detailed analysis demonstrates that the majority of transitions in the non-target regions take place in cells adjacent to the target region. Similar results are obtained when targeting other regions.

5.2 Trojan Detection and Isolation Flow

With large N, each region consists of a small number of components. A small region may magnify the impact of Trojan activity on circuit power consumption because there is less activity in the entire circuit. On the other hand, small regions can decrease Trojan circuit activity, since some of the inputs may be supplied by regions not activated. In contrast, large regions can increase the probability of generating transition inside the Trojan circuit, but the Trojan's impact can be lessened due to an increase in switching activity in the main circuit.

A Trojan can be inserted in any dead space in the circuit layout, and Fig. 5.5 shows the proposed Trojan detection and isolation flow. Given the number of scan chains (No. SCs $= M$), the entire layout is considered as one region (No. Regs $= 1$) and activated by applying random "1"s and "0"s to all scan chains. The power consumption of DUA is measured and compared with that of the golden model. Any measurable difference may indicate the existence of a Trojan; if no Trojan is found, each region can then be divided into smaller regions (the No. Regs must also be updated) to enhance Trojan detection resolution if the region is covered by several scan chains (i.e. No. Regs < No. SCs). The smallest region consists of only one scan chain, and the flow can be continued until the No. Regs is small than the No. SCs. If no measurable power difference is observed in any region at the end of flow, the circuit is considered Trojan free.

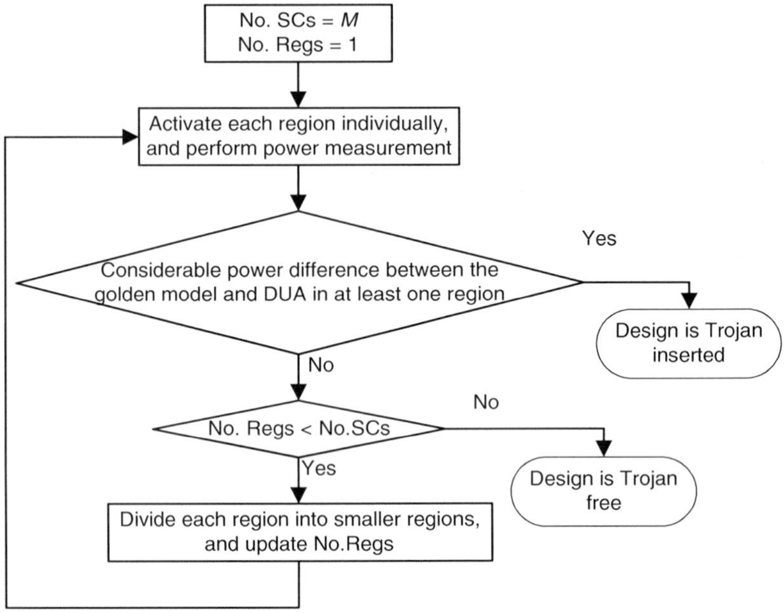

Fig. 5.5 The hardware Trojan detection and isolation flow

A Trojan's cells can be distributed across the entire circuit layout, and their inputs may originate from different regions. Authentication of all combinations of regional activation within circuits with a large N is time consuming. Suppose the circuit is divided into N regions; then there are $\left(\binom{N}{1} + \cdots + \binom{N}{N}\right)$ combinations to be considered during authentication. However, in practice, it is not necessary to examine every combination of regions. The number of gate inputs (i.e. fan-in) is limited by the used technology library. An adversary may not necessarily be able to design a gate with high fan-in to reduce its activity since such a gate greatly impacts the delay characteristics of the circuit and can easily be detected using delay-based techniques [3, 4]. This fact means that only the required inspection is of $\left(\binom{N}{1} + \cdots + \binom{N}{I_{max}}\right)$ combinations where I_{max} is the largest fan-in in the technology library.

Assume that a large circuit is divided into N=48 regions and a tester with a clock frequency of 250 MHz is used. If $I_{max} = 4$, at the first step, every region is examined individually, which needs 48 runs. Then, at the second step, every two regions are inspected, which requires $\binom{48}{2}$ runs. At the third step, examining every three regions requires $\binom{48}{3}$ runs, and finally, $\binom{48}{4}$ runs are needed when targeting every four regions simultaneously. Therefore, a total of 213,052 runs are required (i.e. the maximum number of runs). Suppose $K = 4,000$ test vectors are needed to obtain a measurable power difference between Trojan-inserted and Trojan-free circuits, $4,000 \times \frac{10^{-6}}{250} \times 213,052 = 3.34$ s are required for applying these vectors to the IUA considering all combinations of regions and assuming that in each clock cycle one vector is applied.

If the TE signal is applied differently, as explained in Sect. 5.4, then the application time can increase. After each step, the collected data is analyzed to see if a Trojan is detected. The entire authentication time includes both pattern application time and data processing. Pattern application time can be calculated as above and data processing time depends on the Trojan detection technique [5–7].

5.3 Switching Activity Localization Analysis

The power consumption of a small Trojan is expected to be negligible compared to IUA. Hence, to improve Trojan detection resolution, TCA and TCP must be increased. TCA is a measurement at the logic level to analyze circuit switching activity, while TCP is a measurement at the transistor level to analyze circuit power consumption.

5.3.1 Localization Impact on Circuit Switching Activity

The power consumption of a CMOS circuit is proportional to its switching activity [9]. The relative impact of a Trojan cell on the overall circuit transient current depends on the number of switching cells. Assuming that the probabilities of having "1" and "0" on $Net\ i$ are P_i1 and P_i0, respectively, the transition probability of $Net\ i$ would be $P_i = P_i1 \times P_i0$. A transition on $Net\ i$, as a random variable T_i, can be modeled using the geometric distribution with parameter P_i. The geometric distribution is a discrete distribution for $n = 0, 1, \cdots$, having the probability function $P_i(n) = P_i \times (1 - P_i)^n$ [17]. The probability function states that after n clock cycles, there would be a transition on $Net\ i$ in the $(n + 1)$th clock cycle. Based on the geometric distribution, there would be, on average, one transition on $Net\ i$ in each $(P_i^{-1} - 1)$ clock cycles.

Considering that a circuit consists of N nets, there are $T_1, \cdots, T_N$ random variables. There is one transition in the circuit when there is at least one net that switches. The time between every two consecutive transitions in the circuit is close to the net, which has the minimum T. Simulation results in Sect. 5.1 show that transitions can significantly be limited to a particular region. This indicates that nets in different regions can be considered independent. Therefore, in this analysis, it is assumed that the nets' transitions are independent geometric distribution variables with possibly different parameters P_i, and a transition in the circuit can be stated as

$$T_{Design} = MIN\ T_i \text{ where } 1 \leq i \leq N \tag{5.1}$$

T_{Design} has geometric distribution with parameter P_{Design} given by

$$P_{Design} = 1 - \prod_{i=1}^{N} (1 - P_i) \tag{5.2}$$

It is assumed that nets inside regions are also independent. The correlation is a complex problem and central to any circuit analysis based on a statistical representation of signals, and it can usually be taken care of by using heuristics.

As an example, consider a circuit with 4 nets: Net 1 with the parameter $P_1 = \frac{1}{5}$, Net 2 with $P_2 = \frac{1}{4}$, Net 3 with $P_3 = \frac{2}{3}$, and Net 4 with $P_4 = \frac{1}{6}$. Hence, $P_{Design} = \frac{5}{6}$, and there is one transition in the circuit after each $((\frac{5}{6})^{-1} - 1) = 0.2$ clock cycle on average. Therefore, it is expected that in each clock cycle, the circuit experiences $\frac{1}{0.2}$ transitions, which is called transition density. Transition density is defined as the number of transitions occurring in a target region during each clock cycle.

To evaluate the impact of transition density on Trojan contribution to circuit power consumption, suppose the circuit is divided into two regions, such that there are two nets in each region: Net 1 and Net 2 are in region $R1$, and Net 3 and Net 4 in region $R2$. Further, only one region is active at a time; the other one is kept idle. In this way, the probability parameter of region $R1$ (P_{R1}) is $1-((1-\frac{1}{5})\times(1-\frac{1}{4})) = \frac{2}{5}$. Therefore, on average, every $((\frac{2}{5})^{-1} - 1) = 1.5$ clock cycles, there is one transition in the region $R1$. The transition density of region $R1$ is $\frac{1}{1.5}$, Which, in comparison with transition density on the entire design, $\frac{1}{0.2}$, is about 7.5 times less. This means that if a Trojan is in region $R1$, one transition inside the Trojan is manifested among about 7.5 times fewer transitions compared to when the entire circuit is activated. Therefore, Trojan impact can be significantly magnified if just the region $R1$ is activated instead of the entire circuit. In the same manner, for the region $R2$ the probability parameter of region $R2$ (P_{R2}) is $\frac{13}{18}$, and in each 0.38 clock cycle there is one transition in the region $R2$. The transition density of region $R2$ is $\frac{1}{0.38}$, which is about 2 times less than activating the entire circuit. Therefore, if a Trojan is inserted into region $R2$, its impact can be further manificed by activating just $R2$ than by activating the entire circuit.

In summary, localizing switching activity to a specific region would increase Trojan impact, which in turn increases the sensitivity of current/charge-based detection techniques.

5.3.2 *Localization Impact on Trojan Power Consumption*

In a Trojan-inserted circuit, additional power consumption is expected, relative to the original circuit. To detect Trojans using power-analysis-based methods, Trojan power consumption must be a measurable portion of the circuit power consumption, even in the presence of variations and measurement noise.

Assume that a circuit consists of M components and K power ports. Each component of circuit impacts the power ports in proportion to their distance from them, i.e., the closer the component is to a power port, the more impact it will have on the power port [18]. Therefore, the power measured from the power port k (PPk) can be stated as

$$P_{PPk-trojan-free} = \sum_{1}^{M} d_i P_{comp\#i} \tag{5.3}$$

where d_i denotes the distance of component i from PPk, and $P_{comp\#i}$ is the amount of power consumed by the ith component.

Any extra cell belonging to a Trojan impacts the power at the port k and increases P_{PPk}. When the entire circuit is activated, the ratio of Trojan-to-circuit power consumption (TCP) at power port k (PPk) is

$$TCP_{PPk} = \frac{d_{trojan} P_{trojan}}{\sum_{1}^{M} d_i P_{comp\#i} + d_{trojan} P_{trojan}} \tag{5.4}$$

TCP_{PPk} would be negligible when either a Trojan is far from PPk or when many components are active at the same time. Therefore, it is expected that TCP is increased by activating only a small part of the circuit. If a circuit is divided into N regions, the Trojan-to-circuit power consumption at PPk will change to

$$TCP_{PPk} = \frac{d_{trojan} P_{trojan}}{\sum_{1}^{N} d_{region\#n} P_{region\#n} + d_{trojan} P_{trojan}} \tag{5.5}$$

where $d_{region\#n}$ is the average distance of region n from PPk, $d_{region\#n}$ being a function of the distance of components located in region n, and $P_{region\#n}$ is the amount of power consumed by the nth region. Here, the assumption is that there is no information available about Trojan power consumption since the size and type of Trojans are unknown. Generating transitions in a target region and keeping other regions idle can increase Trojan-to-circuit power consumption ratio since it reduces the amount of circuit power consumption. In other words, since only a few out of N regions are active during power-based analysis, TCP_{PPk} would significantly increase.

A Trojan comparator with 18 inputs is distributed across Regions 2 and 3 of the layout of s38417 used in Sect. 5.1. Two Trojan-inserted circuits are generated to evaluate the effectiveness of switching localization on Trojan detection resolution and to compare with the traditional pattern application: (1) *Circuit A* without reordered scan-cells and (2) *Circuit B* with reordered scan-cells. In the original circuit, *Circuit A*, all scan chains are fed randomly at the same time. However, for *Circuit B*, Region 3 is only activated by applying random patterns to the scan chains covering the region. In each circuit, four power ports are placed at the four corners of the circuit layout. For each circuit, circuit power consumption is measured as the superposition of current drawn from the power ports. The results show that the average and maximum TCPs for *Circuit A* are 0.0025 per vector/clock and 0.21 over the simulation time, while for *Circuit B* the values are 0.0079 and 0.76, respectively. Therefore, switching localization magnifies the average and peak TCPs for s38417 by 3.1X and 3.6X, respectively.

Circuit switching activity would be reduced by applying random "1"s and "0"s to one or more number of scan chains before scan cell reordering. However, the proposed scan cell reordering technique ensures Trojan activation. It is expected that Trojan inputs are supplied by gates located close to Trojan gates unless long wiring from one corner of circuit to another considerably impacts circuit delay characteristics. Therefore, the scan cell reordering technique ensures Trojan activation and localization, The results in Sect. 5.4 support this contention.

5.3.3 Localization Impact on Process Variations

Process variations have raised serious concerns in nanometer regime. Parameters such as device geometry, dopant density, threshold voltage, channel length, and oxide thickness determine circuit delay characteristics and power profile. To study the impact of localization on the contribution of process variations into circuit power consumption and to compare with Trojan contribution, s838 benchmark is divided into four regions, as shown in Fig. 5.3, and a 3-bit counter Trojan is inserted in Region 2. The 180nm process technology parameters are used for Hspice simulations. Monte Carlo (MC) simulation is used to evaluate the impact of process variations modeled with Gaussian distribution in transistor gate length (L) and threshold voltage V_{TH} on the circuit's average power consumption.

To account for inter-die and intra-die process variations in MC simulation, quad-tree based process variations modeling [19] is integrated with MC simulation. Considering both inter-die and intra-die variations for a process parameter P, the value of P can be described as

$$\begin{aligned} P &= P_{nom} + \Delta P_{inter} + \Delta P_{intra}(x_i, y_i) \\ &= P_{nom} + \Delta P_{inter} + \Delta P_{spatial}(x_i, y_i) + \Delta P_{random,i}. \end{aligned} \tag{5.6}$$

P_{nom} represents the mean of P across all possible dies. All devices on a die share one variable ΔP_{inter} for inter-die component variations, which represents the variations of all the gates of a particular die from P_{nom}. $\Delta P_{intra}(x_i, y_i)$ represents intra-die variations and consists of (1) a spatially correlated component $\Delta P_{spatial}(x_i, y_i)$, which is a function of the location of gate i on the die, i.e. (x_i, y_i), and (2) an independent or so-called random component $\Delta P_{random,i}$. which has no correlation with other devices and is represented as a separate random variable for each device [20].

Table 5.1 The standard deviation of V_{TH} in 180 nm CMOS technology with 1.8 V as supply voltage

	$\Delta V_{TH}(mV)$
$NMOS$	$3.635/\sqrt{M \times W_{eff} \times L_{eff}}$
$PMOS$	$4.432/\sqrt{M \times W_{eff} \times L_{eff}}$

To incorporate spatial correlation with MC simulation, the layout of the die is divided into 4 regions using a 3-level quad-tree partitioning. $3\sigma_{total} = 10\,\%$ as total variations on transistor gate length (L) is applied by assigning $3\sigma_{inter} = 4.5\,\%$ as inter-die variations (ΔP_{inter}), $3\sigma_{spatial} = 5.1\,\%$ as spatial correlation ($\Delta P_{spatial}(x_i, y_i)$), and $3\sigma_{random} = 7.2\,\%$ as random variations ($\Delta P_{random,i}$), while $\sigma^2_{total} = \sigma^2_{inter} + \sigma^2_{spatial} + \sigma^2_{random}$ to comply with the jointly normal distribution of normal and independent inter-die, spatial, and random variations. Due to some limitations in the definition of the processing technology library in Hspice, the value V_{TH}, given by the manufacturer, cannot be modified. Hence, to model the impact of V_{TH} variations in an MOS, a DC voltage source is placed in series with the gate terminal of the device, which has Gaussian distribution with zero mean value and standard deviation, as presented in Table 5.1 for NMOS and PMOS transistors. In Table 5.1, W_{eff} and L_{eff} show the values of channel width and length, respectively, and M is the number of parallel devices.

The relation between switching activity localization and the contributions of process variations and Trojans into the circuit's average power consumption are analyzed. Three activation cases are examined: Case1 where all scan chains are fed with random "0" and "1" values; Case2 where Region 2 and Region 3 are activated; Case3 where only Region 2, in which the 3-bit counter Trojan is inserted, is activated.

The results for Case1 are presented in Fig. 5.6a for 50 MC simulation runs. Figure 5.6a shows the percentage of process variations and Trojan contribution into the design's average power consumption. The results for Case1 showed that the contributions of process variations and the Trojan are 4.10 % and 8.45 % on average, respectively. Figure 5.6b presents the results for Case2. The contributions of both process variations and the Trojan are increased to 5.49 % and 11.95 % by limiting switching activity to Regions 2 and 3. The contribution of the Trojan is increased because of increasing TCP, as discussed in Sect. 5.3.2, by reducing circuit switching activity. The contribution of process variations is also increased due to the spatial correlation of gates located in a region. As the process parameters of gates in a region mostly vary in the same direction due to spatial correlation, it is expected that the impact of process variations became magnified by localization, compared with global activation where process parameters vary in different directions across the layout and might relatively neutralize each other. With comparing Case1 and Case2, the difference between the percentage of the contributions of Trojan and process variations is increased from 4.36 % in Case1 to 6.46 % in Case2. It is expected that by limiting switching activity to Region 2, the contributions of both process variations and the Trojan increased, while their difference was further magnified. Figure 5.6c shows the contributions of process variations and the Trojan

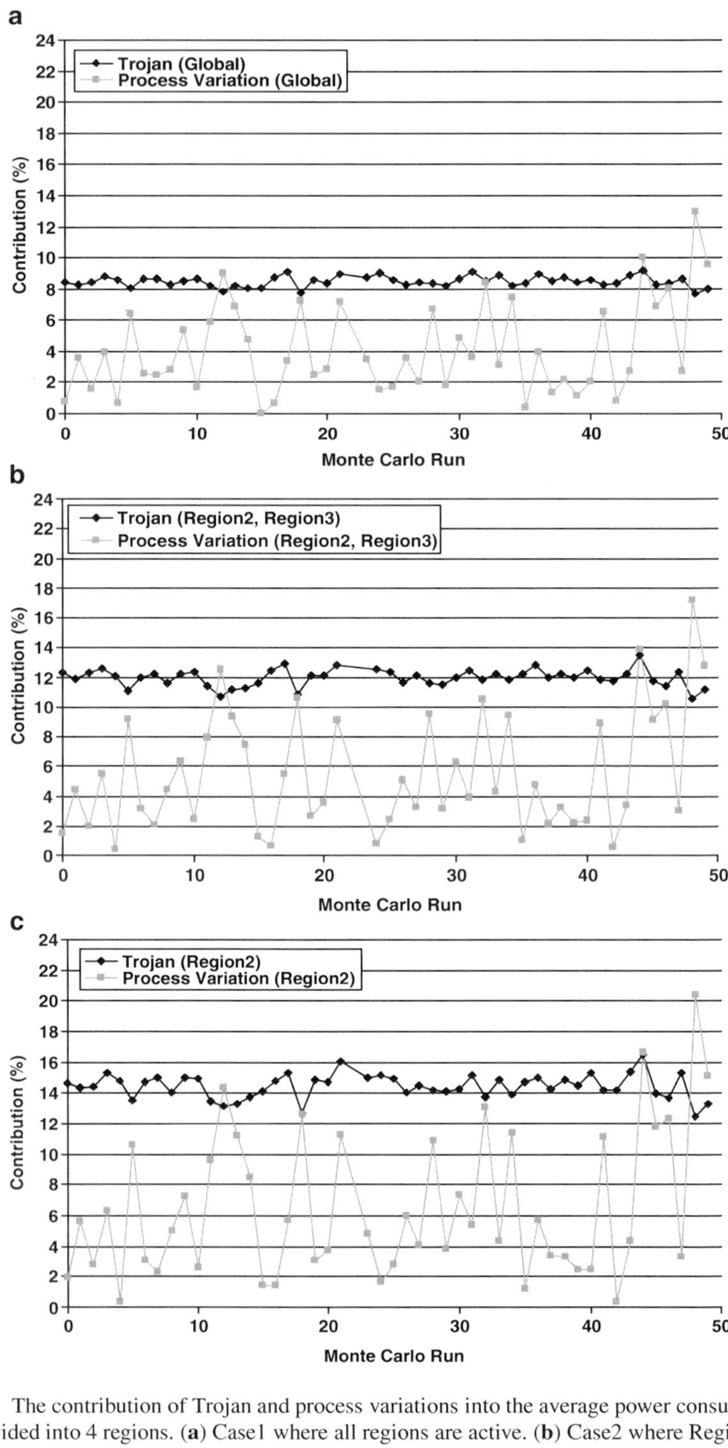

Fig. 5.6 The contribution of Trojan and process variations into the average power consumption of s838 divided into 4 regions. (**a**) Case1 where all regions are active. (**b**) Case2 where Regions 2 and 3 are active. (**c**) Case3 where only Region 2 is active

for Case3. The results show the contribution of process variations and the Trojan increased to 6.53 % and 14.44 %, respectively. Meanwhile, the difference between the contributions of the Trojan and process variations increases to 7.91 %, as expected.

In conclusion, limiting switching activity to a specific region increases the impact of both process variations and Trojans. However, the contribution of Trojan is much higher than that of process variations. Therefore, localization can disclose Trojan impact far beyond process-variations impact.

5.4 Simulation Results

Four same layouts of each s38417 and s35932 benchmarks with a different number of regions are generated (i.e. $N = 4, 9, 16$, and 24.) Four combinational comparator Trojan circuits with 4 (consisting of 4 gates), 6 (consisting of 7 gates), 12 (consisting of 15 gates), 18 (consisting of 23 gates) inputs, named T1, T2, T3, and T4, shown in Fig. 5.7, are designed and inserted into both benchmarks. Furthermore, two sequential Trojans, named T5 and T6, with merely one scan flip-flop, are designed and inserted only into s38417. The clock input of T5 is supplied through a Trojan cone with 4 inputs. The same Trojan cone is used as the data input of T6. For T5, the

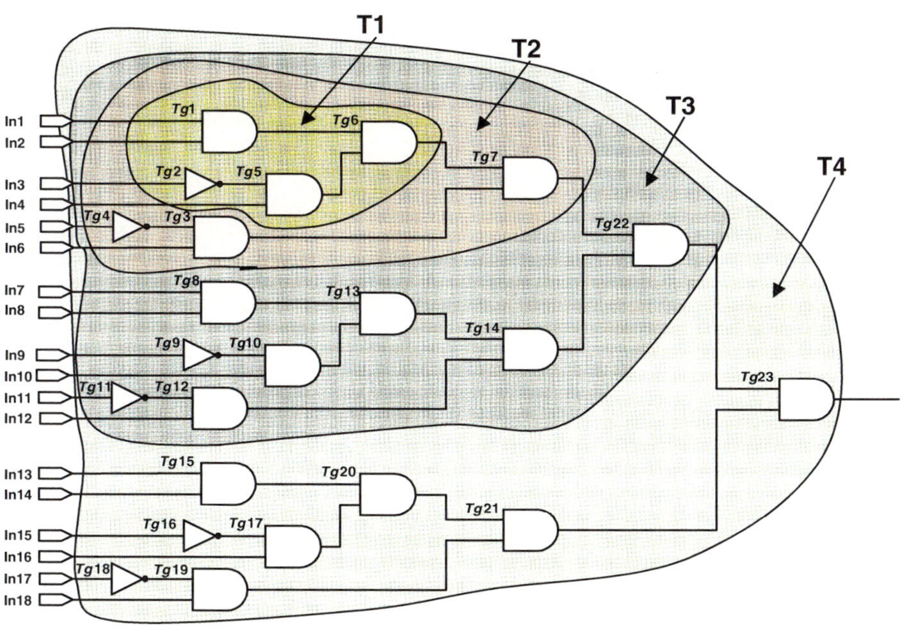

Fig. 5.7 The examined combinational Trojan circuits

Table 5.2 Trojan inputs characteristics

	s38417		s35932	
Input	$P0$	$P1$	$P0$	$P1$
In1	0.56	0.44	0.5	0.5
In2	0.34	0.66	0.5	0.5
In3	0.5	0.5	0.5	0.5
In4	0.5	0.5	0.75	0.25
In5	0.5	0.5	0.5	0.5
In6	0.5	0.5	0.5	0.5
In7	0.5	0.5	0.5	0.5
In8	0.5	0.5	0.75	0.25
In9	0.5	0.5	0.72	0.28
In10	0.5	0.5	0.5	0.5
In11	0.53	0.47	0.25	0.75
In12	0.12	0.88	0.5	0.5
In13	0.5	0.5	0.75	0.25
In14	0.5	0.5	0.5	0.5
In15	0.56	0.43	0.5	0.5
In16	0.42	0.58	0.72	0.28
In17	0.5	0.5	0.5	0.5
In18	0.5	0.5	0.75	0.25
In19	0.9999	10e−4	X	X
In20	0.9999	10e−4	X	X
In21	0.9999	10e−4	X	X
In22	0.9999	10e−4	X	X

Table 5.3 Trojans activation probability

Trojan	S38417	S35932
T1	0.0673	0.030
T2	0.0478	0.0077
T3	6.09e−4	4.39e−05
T4	1.2e−5	9.61e−08
T5:Seq. Trojan 1	Clock activation probability: 9.9e−05	X
T6:Seq. Trojan 2	Data activation probability: 9.9e−05	X

data input of a flip-flop is supplied by a net with the transition probability of 0.25, while the clock of T6 is connected to the circuit's clock. For all Trojans, outputs are left unconnected. Table 5.2 shows the probabilities of "0" ($P0$) and "1" ($P1$) as these Trojans' inputs in both benchmarks. In addition, Table 5.3 indicates the Trojans' activation probabilities.

The original circuits are synthesized using the Synopsys' Design Compiler with a 180nm nanometer technology GSCLib_3.0 library, and physical design is performed by using Synopsys's Astro [16]. The scan-cell reordering procedure is applied with different Ns. After obtaining the DEF file of each new circuit, the circuit is updated by inserting Trojan cells into the circuit's unused spaces and making the required connections. Verilog code corresponding to each circuit is

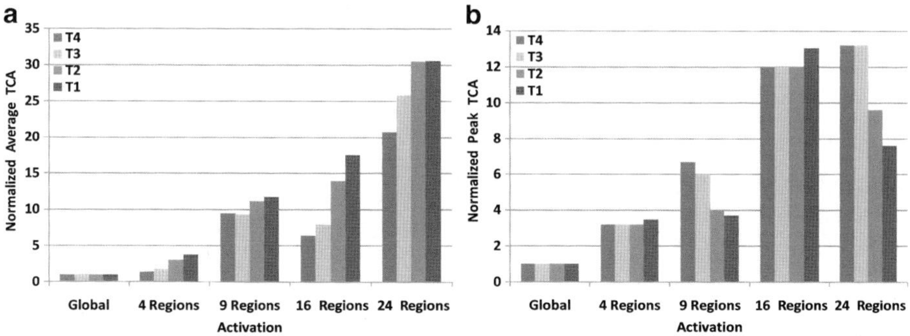

Fig. 5.8 The impact of localized activation on the Average and Peak TCA for s38417

extracted and used for simulation at the logic level [16]. Synopsys' Verilog compiler (VCS) is used to analyze the switching activities of the Trojans and the circuits. Furthermore, the Hspice model of Trojan-inserted circuits is extracted by Synopsys' StarRCXT [16]. Synopsys' NanoSim is used to perform simulation at the circuit level [16]. Four power ports are placed at the four corners of every circuit during the power distribution network synthesis phase, and circuit power consumption is the superposition of current drawn from the power ports.

The term "*local activation*" refers to the use of random patterns in a target region using reordered scan cells and the term "*global activation*" refers to the application of random patterns to the circuit in a traditional manner (i.e. activating all scan chains without reordering scan cells). For each design, simulation is run for $K = 100$ random vectors. A larger number of vectors may be needed for larger circuits.

A Trojan's impact is evaluated by TCA and TCP. Figure 5.8 shows the impact of localization with different Ns on Trojans activity in s38417 benchmark. "*Average TCA*" is defined as the ratio of the total number of transitions inside the Trojan to that of the circuit over the entire simulation time. In Fig. 5.8a, "*Normalized Average TCA*" is the ratio of the average TCA of local activation with specific N to the Average TCA of global activation. Figure 5.8a shows that the Normalized Average TCA for s38417 increases by localizing switching activity to smaller regions (i.e. larger N) by up to 30×. Local activation significantly reduces circuit activity (i.e. background noise) and magnifies Trojan contribution. The results indicate that the Average TCA of global activation is less than local activation for all values of N, even though the larger number of transitions is observed inside the Trojan circuits. For example, the Normalized Average TCA of T4 increases by about 20× with $N = 24$ compared with the global activation in Fig. 5.8a.

"*Peak TCA*" indicates the maximum value of TCA calculated per clock cycle. "*Normalized peak TCA*" is the ratio of the Peak TCA of local activation with specific N to the Peak TCA of global activation in Fig. 5.8b. The Normalized Peak TCA points out cases where Trojan impact exceeds the impact of process variations, and the Trojan can be detected more easily. The results show that even for small Trojans, Normalized Peak TCA increases significantly, which might help effectively

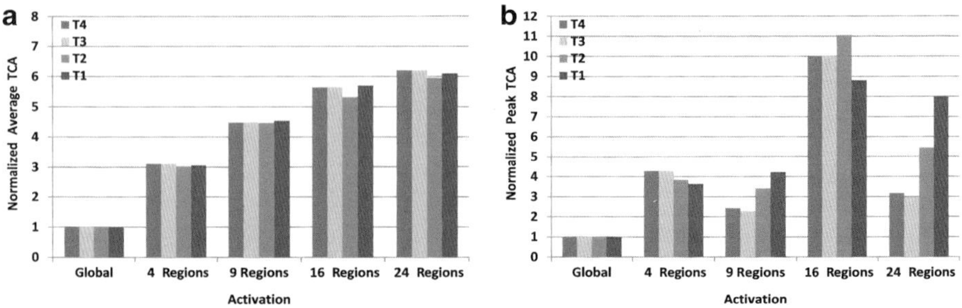

Fig. 5.9 The impact of localized activation on the Average and Peak TCA for s35932

detect them in the presence of process variations. For instance, the Normalized Peak TCA of T4 in local activation with $N = 24$ is about 13× more than that of global activation in Fig. 5.8b. Normalized Average TCA decreased from $N = 9$ to $N = 16$ in T3 and T4, which is attributed to the Trojans' implementation. The Trojans are distributed among several regions, and there are fewer Trojan gates in the target regions. Therefore, the Trojans' activities decreases while the circuits' activity increases with $N = 16$ compared with $N = 9$.

Similar analyses are performed for s35932 benchmark, and results are presents in Fig. 5.9. Figure 5.9a shows that the Normalized Average TCA for s35932 increases with a reduced size of regions (i.e. larger N), and local activation magnifies Trojans activity by about 6× when $N = 24$. The results for the Normalized Peak TCA of s35932 appeared in Fig. 5.9b also shows that Trojan impact can be considerably magnified by local activation. Comparing results for s38417 and s35932 in Figs. 5.8 and 5.9 shows that Trojans in s38417 are more magnified. There are two reasons: (1) Trojans' activation probabilities in s38417 are larger than in s35932 as Table 5.3 illustrates, and (2) circuit activity in s35932 is much higher than in s38417. However, the results still indicated that local activation can considerably magnify Trojan impact, even when the probability of Trojan activation is very low and circuit switching activity is significantly higher.

Figure 5.10 shows TCA per vector in T4 for $N = 1$ (global), 9, and 24. The results for $N = 9$ and 24 are normalized with respect to $N = 1$. The figure emphasizes that the impact of the Trojan on circuit power consumption depends on both Trojan activity and circuit activity. The greater number of transitions in the Trojan does not necessarily make detection easier, as circuit activity may mask Trojan contribution. The results show that most often, in the entire simulation, the TCA of global activation, where $N = 1$, is nearly zero. By contrast, for $N = 9$ and 24, the Trojan contribution is significantly magnified up to 168×. The same analysis is performed for $N = 4$ and 16, and similar results are obtained. Therefore, it is possible to more effectively detect Trojans with local activation when power-based Trojan detection techniques [5–7] are used.

Figure 5.11 shows the contribution (the difference between Trojan-inserted and Trojan-free circuits) of T4 when $N = 24$ on circuit current consumption during

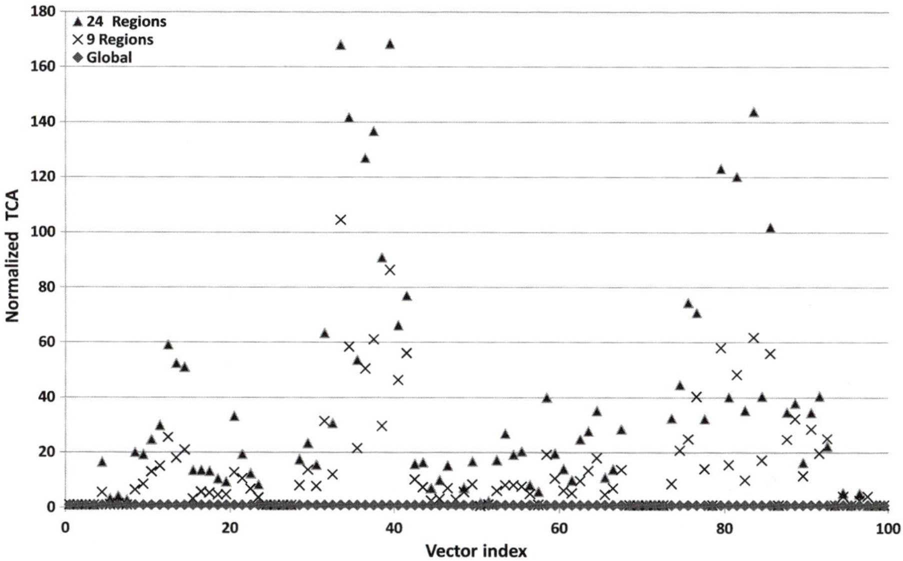

Fig. 5.10 TCA per vector for T4 with $N = 1$ (global), 9, and 24

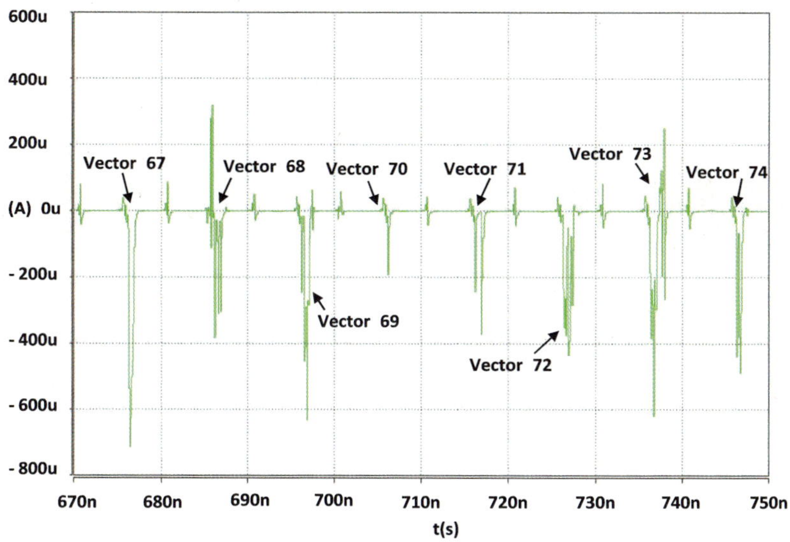

Fig. 5.11 The current consumption of T4 when $N = 24$

the application of vectors 67–74, and Table 5.4 presents the detailed analysis of circuit activity at this interval in local and global activation. For example, in local activation, vector 67 generated one transition in the Trojan (shown as TA) while the number of transitions in the circuit (shown as CA) is 115; therefore, the TCA of local activation would be 0.0087. The same analysis is performed for global activation,

Table 5.4 T4-inserted circuit activity analysis

	Vector index	67	68	69	70	71	72	73	74
Local	TA	1	4	6	1	2	6	8	6
	CA	115	139	135	107	141	105	133	139
	TCA	0.0087	0.028	0.045	0.0093	0.014	0.057	0.060	0.043
	Tj. Avg. Curr. Cons. (μA)	7	4.4	6.2	0.8	2.4	6.2	7.8	4.2
	Cir. Avg. Curr. Cons. (μA)	1,060	1,080	1,100	1,060	1,100	1,120	1,100	1,120
	Current Ratio (Tj./Cir.)	0.0066	0.0040	0.0056	7e−04	0.0021	0.0055	0.0070	0.0037
Global	TA	12	8	5	10	10	6	4	4
	CA	2,051	1,998	1,889	1,763	1,667	1,804	1,788	1,773
	TCA	0.0058	0.0040	0.0026	0.0056	0.0059	0.0033	0.0022	0.0022
	Tj. Avg. Curr. Cons. (μA)	16.6	94	82	7	5.2	26	7.8	8.8
	Cir. Avg. Curr. Cons. (μA)	4,660	4,760	4,800	4,620	4,740	5,200	4,540	4,860
	Current Ratio (Tj./Cir.)	0.0035	0.0019	0.0017	0.0015	0.0010	0.0050	0.0017	0.0018
TCA(Local)/TCA(Global)		1.5	7	17	1.6	2.3	17	27	19

and results show 12 transitions in the Trojan while 2051 transitions are generated inthe circuit, hence, the TCA of global activation would be 0.0058. Comparing both TCAs indicates that global activation masks the Trojan's contribution to total switching because of high switching activity as background noise. The TCA of local activation is 1.6× higher than the TCA of global activation, even though it generates 12 times fewer transitions in the Trojan. Similar results are obtained for other vectors (68–74) as well. The final row in Table 5.4 shows $\frac{TCA(local)}{TCA(global)}$ where in all cases, TCA(local) is considerably larger than TCA(global); up to 27× for vector 73.

The impact of the Trojan on circuit power consumption can be seen in Fig. 5.11. For example, vector 73 has significant impact on the current trace at time 735nsec compared to the other vectors, and Table 5.4 shows that the TCA (and accordingly $\frac{TCA(local)}{TCA(global)}$) of vector 73 is larger than the other vectors. Hspice simulation results show that the average current consumption of the Trojan (Tj. Avg. Curr. Cons.) for vector 73 is 7.8 μA, which is larger than the other vectors. To compare the impact of the Trojan in local and global activations on circuit power consumption, in rows 7 and 13, Trojan power (current) consumption is normalized by the circuit's average current consumption (Cir. Avg. Curr. Cons.) per vector. The normalized values also show the larger impact of the Trojan on circuit power consumption during local activation.

The contributions of sequential Trojans are analyzed by measuring TCA and the average TCPs per positive level of clock cycles. s38417 is divided into four regions, and T5 is inserted into one of regions. TCA measurements show that the average and peak values for local activation are 2.7× and 4.7× larger than those for global activation. Figure 5.12 shows the TCP measurement per vector. The figure shows that TCP for local activation is above TCP for global activation for most vectors. The results show that the Average and Peak TCPs for local activation are 1.3× and 2.4× larger than those for global activation. A similar analysis is performed for T6 and the results show that the Average and Peak TCPs for local activation are 1.3×

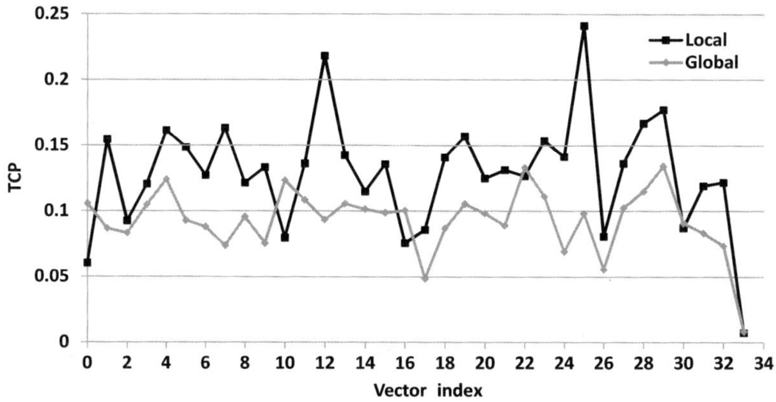

Fig. 5.12 TCP measurement of T5

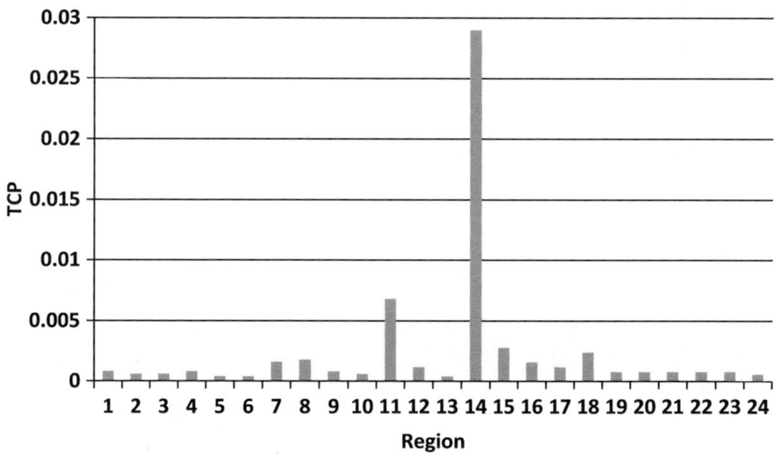

Fig. 5.13 The Average TCP per region

and 3.8× larger than those for global activation. Comparing the results for T5 and T6 shows that the Average TCP was the same for both Trojans. However, T6, connected to clock tree, has greater impact on circuit power consumption and increases Peak TCP.

Local activation can also determine a Trojan's location. In s38417's layout with 24 regions, the Trojan cells of T4 are distributed in regions 14 and 11. All regions are separately activated by applying three different random patterns. For each region, the TCPs per each positive level of clock cycles are measured. Figure 5.13 presents the measurement for the second pattern corresponding to each region. The results clearly show that regions 14 and 11 have the highest average TCP among all regions, which is an indication of Trojan existence in those regions. A more detailed analysis of regions 14 and 11 reveals that the number of transitions inside the Trojan's

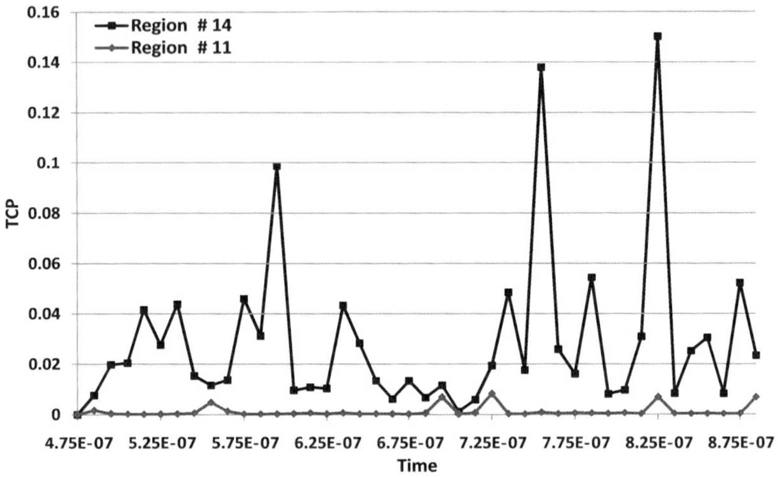

Fig. 5.14 Localization of T4 with TCP measurement

cells inserted into region 14 is more than 2× that of those inserted into region 11. Furthermore, Fig. 5.14 presents TCP for Regions 11 and 14 during the second pattern from 485 to 885nsec. The results indicate that the TCP of region 14 is significantly higher than that of region 11 for a number of clock cycles. Therefore, the average TCP of region 14 is larger than that of region 11, as indicated in Fig. 5.13.

An adversary may design a Trojan circuit to be inactive during authentication when the TE signal is active. A Trojan may use the TE signal as a trigger input and start working when the TE signal is inactive, i.e. the circuit is in the functional mode. To address this issue, the TE signal must be switched on and off frequently. When the TE signal is on (high), the circuit is in shift (or scan) mode, and one or more bits are shifted into the scan chain. To see the impact of a new random pattern in the scan chain, the TE signal is switched back to low. This will allow the pattern to be applied from scan flip-flops to the circuit under testing, and responses will go back to the scan chain.

T4 is equipped with the TE signal such that it is fully functional only when the TE signal is off. That is, T4 functions as a comparator when the TE signal is "0"; otherwise, it is only subjected to partial activation. In Fig. 5.7 all gates connected to T4's inputs are controlled by the TE signal except $Tg1$, $Tg11$, $Tg12$, $Tg18$, and $Tg19$. To evaluate the alternation of the TE signal, four cases are simulated and the Normalized Average and Peak TCAs are measured for $N = 9$ and 24. In the first case, TE1(1)0(1), the TE signal is switched by each clock cycle. In the second case, TE1(10)0(1), the TE signal is on for ten clock cycles and then switched off for one clock cycle. In the third case, TE1(30)0(1), the on state of the TE signal lasts for 30 clock cycles and then switches off for one clock cycle. The final case was TE1(100)0(0) where the TE signal is kept high for the entire simulation.

Table 5.5 Test enable signal alteration analysis for T4

	Normalized average TCA		Normalized peak TCA	
	9 Regions	24 Regions	9 Regions	24 Regions
TE1(1)0(1)	1.29×	1.57×	1.46×	2.68×
TE1(10)0(1)	2.4×	2.6×	2.31×	2.75×
TE1(30)0(1)	3.59×	4.26×	5.90×	5.68×
TE1(100)0(0)	6.19×	11.90×	5.55×	16.66×

Lengthening the high duration of the TE signal would magnify TCA since the circuit is subjected to less switching activity. The results, presented in Table 5.5, show that T4 impact would be exposed by switching the TE signal between on and off. The results show that the Normalized Average and Peak TCAs considerably increase (up to 12×) by using larger N and keeping the TE signal high for a greater number of clock cycles. It should be noted that the Normalized Average and Peak TCAs are lower compared with results shown in Fig. 5.8. This is attributed to functional dependency among the vectors when the TE signal switches on and off.

5.5 Summary

A new layout-aware scan-cell reordering method was presented, aiming at limiting switching activity to a specific region to improve Trojan detection. Two new metrics, namely Trojan-to-circuit activity (TCA) and Trojan-to-circuit power consumption (TCP), were introduced to measure the effectiveness of Trojan detection techniques. Switching localization impacts on circuit switching activity, Trojan power consumption, and process variations were analyzed statistically and by simulation. The results showed that switching in most of the non-target regions can be reduced significantly. The impact of the region's size was evaluated, and the results indicated that smaller regions can more effectively magnify Trojan activity, compared to global activation. Meanwhile, this method can be used to localize hardware Trojans.

References

1. P. C. Kocher, J. Jaffe, and B. Jun, "Differential Power Analysis," in Proc. of the *CRYPTO*, vol. 1666 of Lecture Notes in Computer Science, pp. 388–397.
2. M. Tehranipoor and F. Koushanfar, "A Survey of Hardware Trojan Taxonomy and Detection," in *IEEE Design and Test of Computers*, pp. 10–25, 2010.
3. Y. Jin and Y. Makris, "Hardware Trojan Detection using Path Delay Fingerprint," in Proc. of the *IEEE International Workshop on Hardware-Oriented Security and Trust (HOST2008)*, pp. 51–57, 2008.

4. D. Rai and J. Lach, "Performance of Delay-based Trojan Detection Techniques under Parameter Variations," in Proc. of the *IEEE International Workshop on Hardware-Oriented Security and Trust (HOST2009)*, pp. 58–65, 2009.
5. D. Agrawal, S. Baktir, D. Karakoyunlu, P. Rohatgi, and B. Sunar, "Trojan Detection using IC Fingerprinting," in Proc. of the *Symposium on Security and Privacy*, pp. 296–310, 2007.
6. R. Rad, X. Wang, J. Plusquellic, and M. Tehranipoor, "Taxonomy of Trojans and Methods of Detection for IC Trust," in Proc. of the *International Conference on Computer-Aided Design (ICCAD08)*, pp. 632–639, 2008.
7. X. Wang, H. Salmani, M. Tehranipoor, and J. Plusquellic, "Hardware Trojan Detection and Isolation using Current Integration and Localized Current Analysis," in Proc. of the *International Symposium on Fault and Defect Tolerance in VLSI Systems (DFT08)*, pp. 87–95, 2008.
8. R. Sankaralingam, R.R. Oruganti, and N.A. Touba, "Static Compaction Techniques to Control Scan Vector Power Dissipation," in Proc. of the *IEEE VLSI Test Symposium (VTS00)*, pp. 35–40, 2000.
9. S. Devadas and S. Malik, "A Survey of Optimization Techniques Targeting Low Power VLSI Circuits," in Proc. of the *Design Automation Conference(DAC95)*, pp. 242–247, 1995.
10. N. Badereddine, P. Girard, S. Pravossoudovitch, A. Virazel, and C. Landrault, "Scan Cell Reordering for Peak Power Reduction during Scan Test Cycles," IFIP International Federation for Information Processing, ISBN 978-0-387-73660-0, pp. 267–281, 2007.
11. Y.Z. Wu and M.C.-T. Chao, "Scan-Chain Reordering for Minimizing Scan-Shift Power Based on Non-Specified Test Cubes," in Proc. of the *IEEE VLSI Test Symposium (VTS08)*, pp. 147–154, 2008.
12. W. Li, S. Wang, S.T. Chakradhar, and S.M. Reddy, "Distance Restricted Scan Chain Reordering to Enhance Delay Fault Coverage," in Proc. of the *International Conference on VLSI Design*, pp. 471–478, 2005.
13. N. Devtaprasanna, S.M. Reddy, A. Gunda, P. Krishnamurthy, and I. Pomeranz, "Improved Delay Fault Coverage using Subsets of Flip-flops to Launch Transitions," in Proc. of the *IEEE Asian Test Symposium (ATS05)*, pp. 202–207, 2005.
14. M. Hirech, J. Beausang, and X. Gu, "A New Approach to Scan Chain Reordering using Physical Design Information," in Proc. of the *IEEE International Test Conference 1998 (ITC98)*, pp. 348, 1998.
15. L. Hsu and H. Chen, "On Optimizing Scan Testing Power and Routing Cost in Scan Chain Design," in Proc. of the *International Symposium on Quality Electronic Design (ISQED06)*, pp. 451–456, 2006.
16. Synopsys Inc., User Manual, http://www.synopsys.com/.
17. D.D. Wackerly, W. Mendenhall III, and R.L. Scheaffer, "Mathematical Statistics with Application, 7th edition" Thomson Learning, Inc., 2008.
18. J. Ma, J. Lee, and M. Tehranipoor, "Layout-Aware Pattern Generation for Maximizing Supply Noise Effects on Critical Paths," in Proc. of the *IEEE VLSI Test Symposium (VTS09)*, pp. 221–226, 2009.
19. A. Agarwal, D. Blaauw, V. Zolotov, S. Sundareswaran, M. Zhao, K. Gala, and R. Panda, "Statistical delay computation considering spatial correlations," in Proc. of the *Asia and South Pacific Design Automation Conference (ASP-DAC03)*, pp. 271–276, 2003.
20. A. Srivastava, D. Sylvester, and D. Blaauw "Statistical Analysis and Optimization for VLSI: Timing and Power," Springer, 2005.

Chapter 6
Design for Hardware Trust: Ring Oscillator Network

In this chapter, a hardware Trojan detection method is presented, which is performed by combining measurements from an on-chip structure with external dynamic current measurements [1]. This method monitors power fluctuations and differentiates fluctuations due to hardware Trojans from fluctuations due to measurement noise and process variations. This method considers Trojan impact on the power consumption of neighboring cells and the entire IC. A number of ring oscillators (ROs) acting as *power monitors* are inserted into a circuit, forming a ring oscillator network (RON). Each row of the circuit under authentication contains at least one inverter of an RO in the RON. Thus any malicious inclusions in each row would be captured by one of these ring oscillators. Off-chip test equipment will measure the transient current of the IC, which will be combined with the ROs' cycle counts to generate a power signature for the entire IC. The signature of the CUT is then compared against the Trojan-free signatures.

6.1 Analyzing Impact of Power Supply Noise on Ring Oscillators

Two simple five-stage ring oscillators are shown in Fig. 6.1: the ring oscillator in Fig. 6.1a consists of inverters and the ring oscillator in Fig. 6.1b is composed of NAND gates. The second ring oscillator has a higher sensitivity to supply noise since one of its inputs is connected to the power supply but this ring oscillator offers larger area overhead. The first ring oscillator is used as a power monitor since its behavior can easily be described analytically. The frequency of this ring oscillator is determined by the total delay of all the inverters in the presence of supply voltage variations and process variations. Assume that each stage in the ring oscillator provides a delay of t_d. The delay of the n-stage ring oscillator, then, is approximately $2 \times n \times t_d$ and the oscillation frequency will be $f = 1/(2 \times n \times t_d)$. The delay of each inverter is given by

M. Tehranipoor et al., *Integrated Circuit Authentication: Hardware Trojans and Counterfeit Detection*, DOI 10.1007/978-3-319-00816-5_6,

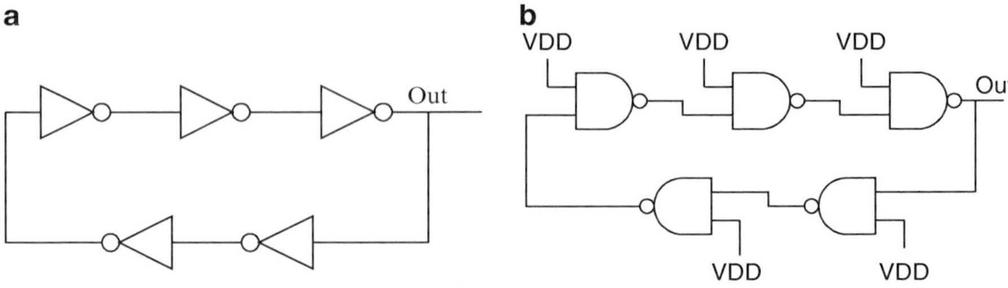

Fig. 6.1 Five-stage ring oscillators

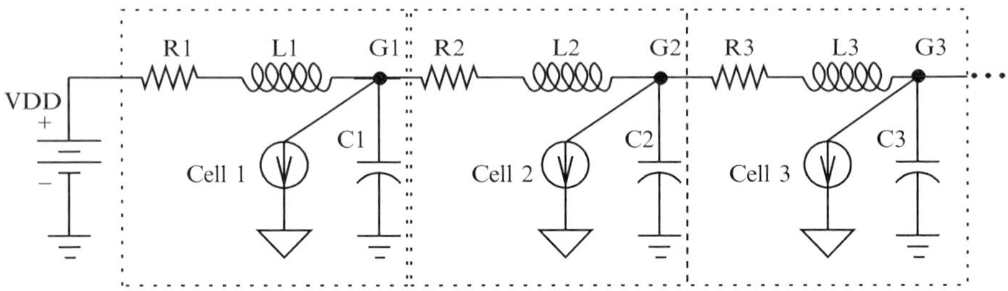

Fig. 6.2 The RLC model of a simple power line in a power distribution network

$$t_d = 0.52 \times \frac{C_L V_{dd}}{(W/L)k' V_{DSAT}(V_{dd} - V_{th} - V_{DSAT}/2)} \tag{6.1}$$

where C_L is load capacitance, k' represents transconductance, and W/L denotes the transistor channel width to channel length ratio. V_{DSAT}, V_{th}, and V_{dd} represent saturation voltage, threshold voltage, and power supply voltage, respectively [2].

Power supply noise (also known as voltage drop) increases the delays of logic gates in an IC. Thus, a change in the supply voltage of any inverter in a ring oscillator impacts the delay of all associated gates, and therefore impacts the oscillation frequency.

Concerning today's tightly designed power supply distribution networks, transitions in one gate can impact the power supply of other gates within close proximity [3]. Figure 6.2 shows a simple power line model in which VDD line supplies one row in the standard cell design. The indicated VDD line represents the point where a via connects the power rail to the upper metal layer in a power distribution network. Nodes G1, G2, and G3 connect to adjacent cells represented as current sources in Cell 1, Cell 2, and Cell 3. Here, for the sake of simplicity, the power via is assumed to have zero impedance and each interconnect is modeled by a resistance, inductance, and capacitance (RLC) network. The contribution of each current source to the overall noise is described in (6.2) where $V1$, $V2$, and $V3$ (voltage at nodes G1, G2, and G3) are the power supply noise spectrum, $Vii = Z_{ii} \times I_{ii}(i = 1, 2, 3)$ (Z_{ii} is the impedance of node i and I_{ii} is the current) is the power noise,

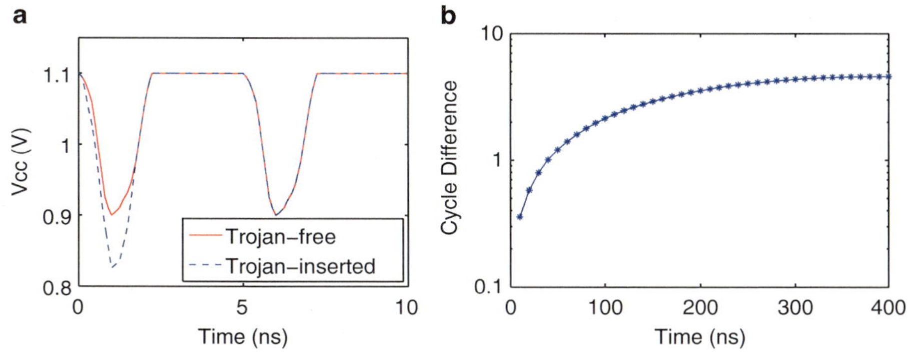

Fig. 6.3 (**a**) Power supply variations for Trojan-free and Trojan-inserted circuits; (**b**) Cycle difference caused by Trojan gates' switching

$\rho_{ij}(i, j = 1, 2, 3)$ is the voltage division coefficient, and ω is the frequency of the circuit. As can be seen from the equation, $V1$, $V2$, and $V3$ are related to the power noise of neighboring gates, demonstrating that a transitioning gate has an effect on neighboring gates connected to the same VDD line.

$$\begin{aligned} V1 &= V11 + \rho_{21}(\omega) \times V22 + \rho_{31}(\omega) \times V33 \\ V2 &= \rho_{12}(\omega) \times V11 + V22 + \rho_{32}(\omega) \times V33 \\ V3 &= \rho_{13}(\omega) \times V11 + \rho_{23}(\omega) \times V22 + V33 \end{aligned} \tag{6.2}$$

Therefore, with the same input patterns, the power supply noise affecting the Trojan-free IC and Trojan-inserted IC will differ due to the switching gates within a Trojan. In order to verify the impact of the Trojan on the frequency of the ring oscillator, a 5-stage ring oscillator (shown in Fig. 6.1a) is implemented in the 90 nm technology for simulation.

In Fig. 6.3a, the dashed line denotes the dynamic power in the presence of a Trojan and the solid line denotes the Trojan-free power (assuming $VDD = 1.2$ V). The two supply voltages only differ during the first 2 ns. These two power waveforms are applied to the ring oscillator for 400 ns. Figure 6.3b shows the cycle count difference between the Trojan-free and Trojan-inserted ICs. At time 0, the two ring oscillators denoting *with* and *without* an inserted Trojan have the same period. However, due to the Trojan-induced power supply noise, the cycle count difference grows steadily as the measurement duration increases.

A Trojan, composed of 20 combinational gates, is also simulated to demonstrate its effect on the frequency of a 5-stage ring oscillator at 25°C. The simulation time is 10 μs. Figure 6.4a shows the locations of the ring oscillator and the Trojan: the ring oscillator is placed at the left corner of a standard cell row, while the Trojan is located at some position between locations *1* and *11*. There is one flip-flop (FF) between every two possible Trojan locations. CC_f denotes the cycle count of the

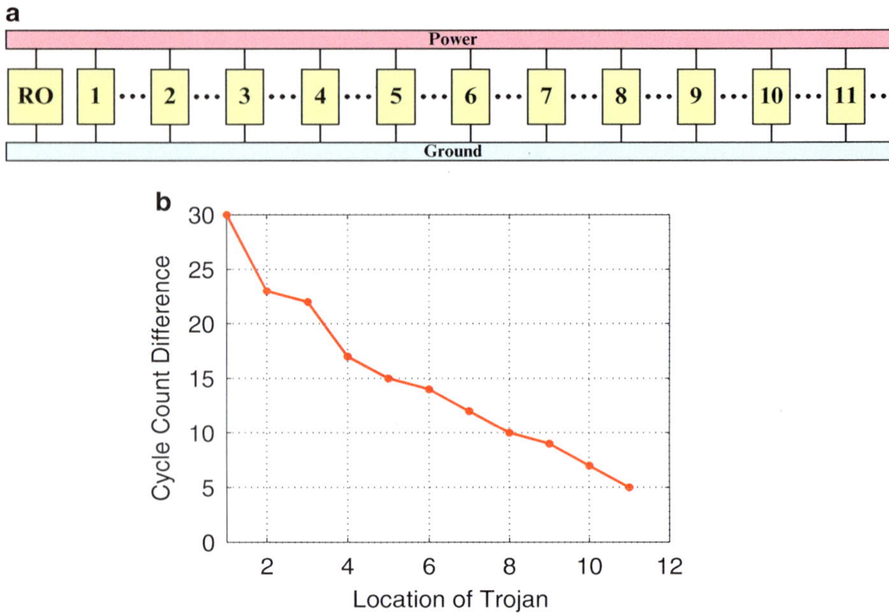

Fig. 6.4 (**a**) RO and Trojan's location; (**b**) Cycle count difference caused by Trojan gates-induced voltage drop

ring oscillator in a Trojan-free IC, CC_t denotes the cycle count of the ring oscillator in an IC with a Trojan, and $\Delta CC_{ft} = CC_f - CC_t$. Figure 6.4b illustrates the relationship between ΔCC_{ft} and the Trojan's location. As can be seen in the figure, the Trojan gates' switching will reduce the ring oscillator's frequency by an amount related to the distance between the Trojan and the ring oscillator. The farther the Trojan is placed from the ring oscillator, the less impact it has on the ring oscillator's frequency.

6.2 The Relationship Between RO Frequency and Localized and Total Dynamic Current

The delay of each inverter in the ring oscillator can also be expressed as $t_d = k_g/I_g$ where k_g is a gate-dependent constant, and I_g is the dynamic current of the inverter [6]. Based on the Alpha-Power Model mentioned in [7], the dynamic current of a switching gate is

$$I = \mu_g \times (V_{dd} - V_{th})^{\alpha} \tag{6.3}$$

where α is the velocity saturation index. Thus the frequency of the n-stage ring oscillator can be expressed as:

$$f = \frac{\mu_g \times (V_{dd} - V_{th})^{\alpha}}{2n \times k_g} \tag{6.4}$$

In the presence of a Trojan, the ring oscillator frequency is modeled by (6.5) rather than (6.4) where the voltage-drop ΔV_t represents the Trojan's contribution to the ring oscillator frequency. As the equation shows, the frequency of the ring oscillator f_t is more sensitive to the voltage-drop ΔV_t when the stage of the ring oscillator n is smaller. However, if n is too small, the frequency of the ring oscillator will be too high to be measured by an on-chip counter. Using (i) an operating frequency of f=1 GHz, (ii) $V_{dd} = 1.2$ V, and (iii) Synopsys 90nm technology in a Nanosim simulation, a 5-stage RO would be the smallest allowable RO. Thus, 5-stage ring oscillators will be used.

$$f_t = \frac{\mu_g \times (V_{dd} - \Delta V_t - V_{th})^{\alpha}}{2n \times k_g} \tag{6.5}$$

The dynamic current of the entire Trojan-free chip is:

$$I_{total} = \sum_{i=0}^{i=N} \lambda_i \times N \times \mu_g \times (V_{dd} - V_{th})^{\alpha} \tag{6.6}$$

where N is the total number of switching gates in the IC, and λ_i denotes the gate-dependent constant of the i_{th} gate. The constant, λ_i, depends only on the type of gate specified, not the particular instance of such a gate. The relationship between the frequency of the n-stage ring oscillator embedded into the chip and the dynamic current of entire chip will be:

$$\frac{I_{total}}{f} = \sum_{i=0}^{i=N} \lambda_i \times N \times 2n \times k_g \tag{6.7}$$

For ICs with n_t Trojan gates inserted, (6.7) becomes:

$$\frac{I_{total,t}}{f_t} \approx \sum_{i=0}^{i=N+n_t} \lambda_i \times (N + n_t) \times 2n \times k_g (1 + \alpha \times \frac{\Delta V_t}{V_{dd} - \Delta V_t - V_{th}}) \tag{6.8}$$

when $\Delta V_t << V_{dd} - \Delta V_t - V_{th}$ [9]. By comparing (6.8) with (6.5), it can be concluded that combining ring oscillator frequency measurements with current measurements will achieve greater sensitivity to Trojans than either measurement alone.

The above analysis is based on ring oscillators made with standard threshold voltage (SVT) transistors. However, ring oscillators with high threshold voltage (HVT) transistors are more sensitive to power supply noise, as shown by the simulation results in Fig. 6.5, performed using the same technology with a 5-stage

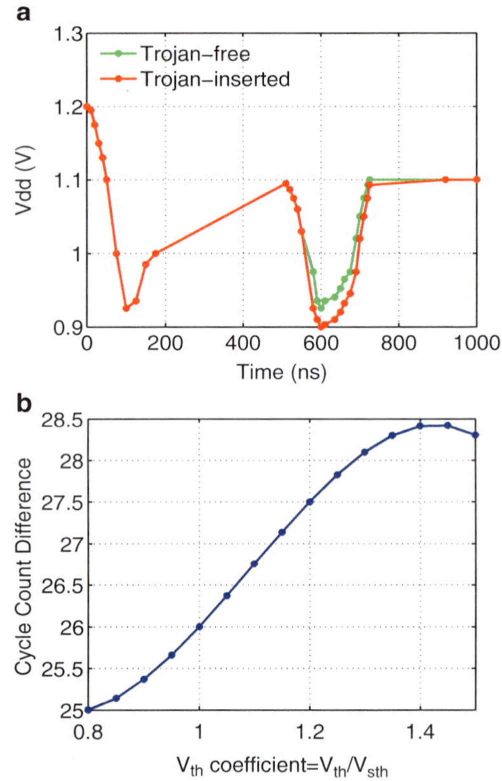

Fig. 6.5 (**a**) Power supply variations for Trojan-free and Trojan-inserted circuits; (**b**) Cycle count difference increases as threshold voltage increases

ring oscillator. The green line in Fig. 6.5a denotes the power supply voltage of Trojan-free ICs during the 1,000 ns simulation period while the red line represents the power supply voltage of Trojan-inserted ICs. Figure 6.5b shows that for a particular ring oscillator, the cycle count difference between Trojan-free ICs and Trojan-inserted ICs will increase as the threshold voltage of the transistors increases until a maximum is reached. Once this maximum has been reached, increasing the threshold adversely affects the cycle-count difference (and thus the sensitivity to inserted Trojans). The X axis in Fig. 6.5b represents the threshold voltage coefficient, V_{th}/V_{sth}, where V_{sth} is the SVT of the MOS transistors. In the Synopsys 90nm technology library, the threshold voltage coefficient of the HVT transistors is 1.2. With HVT ring oscillators, (6.8) becomes:

$$\frac{I_{total,t}}{f_t} = \sum_{i=0}^{i=N+n_t} \lambda_i \times (N + n_t) \times 2n \times k_g (1 + \alpha \times \frac{\Delta V_t + V_{hth} - V_{sth}}{V_{dd} - \Delta V_t - V_{hth}}) \quad (6.9)$$

where V_{hth} is the high threshold voltage of the transistors in the ring oscillators. As seen in (6.9), the relationship between the IC's dynamic current and the frequency of a ring oscillator in the circuit will be more sensitive using HVT transistors. In addition, Trojans with larger (n_t) and more IR-drop (ΔV_t) are easier to detect.

Some of the parameters in (6.9) will change with process and environmental variations. Since ICs are tested under the same temperature condition in a production test environment, only small environmental variations will be considered. All remaining parameters are susceptible to process variations and statistical analysis will help to separate the contributions of process variations and Trojans to the transient power.

6.3 Ring Oscillator Network Structure

As previously discussed, Trojan gates' switching impacts both the frequency of nearby ring oscillators and the IC's dynamic current. Since a Trojan's effects may be localized (i.e. tightly distributed), and the impact of a Trojan on a ring oscillator is dependent upon the distance between them, one ring oscillator may not be sensitive enough to distinguish the effects of Trojans from process variations throughout the entire IC. An improved RON, however, can improve sensitivity to Trojan noise.

Figure 6.6 shows a circuit into which the proposed on-chip structure with N_{RO} n-stage ring oscillators is inserted. These n-stage ring oscillators are each composed of one NAND gate and $n-1$ inverters with one component located in each of the n rows of the standard cell design. The ring oscillators are more sensitive to the voltage drop caused by a Trojan if they share the same power strap. Therefore, it is highly advantageous to ensure complete coverage of the power distribution network by placing at least one ring oscillator component in each row of the standard cell (and thus near each power strap). One set of n-stage ring oscillators will be inserted between two vertical straps. If there are M vertical power straps and R rows in the design, $N_{RO} = (M+1)\lceil R/n \rceil$. However, the number of ring oscillators can be adjusted according to the required Trojan detection sensitivity and the minimum sensitivity to Trojan activity.

The on-chip structure also includes a linear feedback shift register (LFSR), one decoder, one multiplexer, and one counter. The LFSR will supply random functional patterns for the entire IC during the signature generation and authentication processes; the same seed must be used for each golden IC and each IC under authentication. The decoder and the multiplexer are used to select which ring oscillator is measured. When a ring oscillator is selected, the decoder enables that particular RO and the multiplexer transmits the output of that RO to the counter. The counter measures the cycle count of the selected ring oscillator over a specified duration. The number of stages in a ring oscillator is limited by the operating speed of the counter, which is determined by the technology node. For example, using Synopsys 90nm technology, a 16-bit counter can operate at a maximum frequency of $1GHz$ according to HSPICE simulation.

Since the ring oscillators are only enabled during the production test and the authentication phase, their power overhead in the field is negligible. The proposed architecture has a small area overhead, due mainly to the ring oscillators. For larger

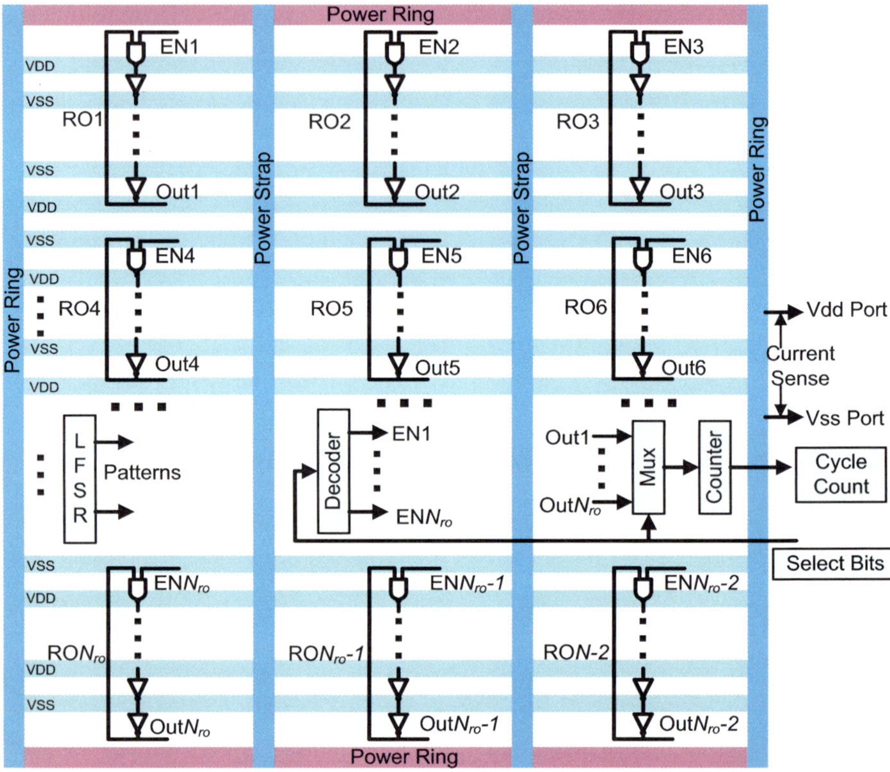

Fig. 6.6 The proposed on-chip structure with each gate of the ring oscillators placed in a standard cell row

circuits, assuming there is one vertical power strap for every 20 FFs or 80 gates, the area overhead of the ring oscillators will be approximately $1/(20 \times 4) = 1.25\,\%$. The total area overhead will be approximately 2.5% if there is only one vertical strap for every 10 FFs or 40 gates in the design. For a small circuit, the counter may play a significant part in the area overhead, but the counter size does not increase linearly with the size of the circuit. Since LFSRs are commonly used for built-in self-tests in modern designs, it can be ignored when analyzing the area overhead. However, even with LFSRs, the area overhead of a RON in large designs is still quiet small, since the area of the LFSRs does not increase significantly with the size of the circuit either. Transient current will be measured externally (i.e. with no area cost). In summary, the area overhead will be less than 3% for a large circuit and would be slightly larger for a smaller circuit. For instance, the overhead of RONs with LFSRs is 5.58% for the ISCAS'89 benchmark circuit s38584 (which contains four vertical power straps), 2.47% for an AES circuit (with six vertical straps), and 1.99% for a DES circuit (with six vertical power straps). The AES and DES circuits are provided in [10].

Since the ring oscillators are distributed across the entire IC, it is inherently difficult for an adversary to remove or tamper with one. If one of the ring oscillators reports data outside of a certain range or does not report data, it must have been attacked. In addition, this proposed on-chip structure is resilient to modeling. Some attackers may build up a lookup table to repeatedly generate the same cycle count for each ring oscillator, which would attempt to replace the Trojan-effected counter values with known good values. However, the current consumed by the lookup tables may be captured by the external current measurement and the power signature generated by the outlier analysis would also be changed. On the other hand, if the ROs are replaced with lookup tables embedded in the design, the frequency of the same ring oscillator at the same location, but on a different chip would stay at the same value in different ICs. However, unlike the value stored in a lookup table, the measured frequency of an RO in different ICs should be slightly different due to different process variations in Trojan-free circuits. If one ring oscillator in all CUTs has exactly the same frequency, designers will easily know that the IC was tampered with using embedded lookup tables.

6.4 Measurement Flow and Statistical Analysis

The measurement flow for each IC is shown in Fig. 6.7. To measure the frequency of N_{RO} ring oscillators, the LSFR patterns with the same seed will be applied N_{RO} times. The transient current is measured externally. A signature is generated by recording the cycle count of each ring oscillator and the transient current from a large number of ICs of the same design. Since the ICs will all be subject to different process variations, this signature can be statistically more tolerant to similar variations in chips under authentication. In order to separate the effect of process variations and Trojans, a data analysis flow is suggested which includes the following three methods: (i) Simple Outlier Analysis, (ii) Principal Component Analysis (PCA), and (iii) Advanced Outlier Analysis.

Simple outlier analysis is based on the statistical distribution of the oscillation cycle count for each ring oscillator in the RON. For each ring oscillator, the oscillation count is within a certain range for Trojan-free ICs. If the oscillation count of even one ring oscillator in the IC under authentication is outside of the range, this IC is considered suspicious and might contain a Trojan. This method uses the information from individual ring oscillators but not the relationships among ring oscillators in the network, nor the dynamic current of the entire IC. This method can often identify a small number of Trojan-inserted ICs but may fail to detect most Trojan-inserted ICs. If the oscillation cycle count of all ring oscillators in an IC under authentication is within each Trojan-free IC's signature, the data collected from this IC will be processed by PCA and advanced outlier analysis.

The principal component analysis method [4] is used to account for the $N_{RO} + 1$ variables. With one variable representing one ring oscillator, there are N_{RO} variables, and the $N_{RO} + 1$th variable represents the dynamic current. The relationship

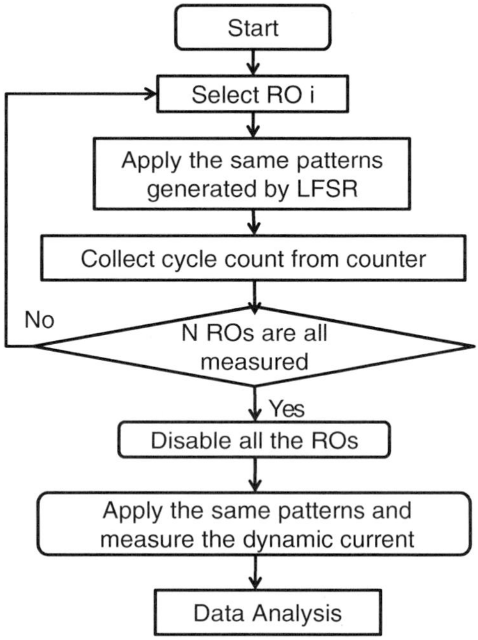

Fig. 6.7 Measurement flow of proposed method

between the data from the N_{RO} ring oscillators and the dynamic current is considered by PCA when it transforms the $N_{RO} + 1$ variables into uncorrelated variables. The $N_{RO} + 1$ variables are transformed by PCA and the first three of the resulting components in Trojan-free ICs are used to construct a convex hull [5]. If the output of the CUT is beyond the convex, a Trojan must exist in the IC under authentication. However, if the output is inside the convex, an advanced outlier will be used for further analysis and validation.

An advanced outlier analysis has been developed to identify Trojan-inserted ICs that cannot be detected by simple outlier analysis and PCA. It considers the relationships among ROs in the RON and the dynamic current of the entire chip. The pseudo-code is shown in Fig. 6.8. For each Trojan-free IC, two out of N_{RO} ring oscillators will be selected along with the dynamic current information to generate a power signature (shown in Fig. 6.8a). For a particular Trojan-free IC, the total oscillation cycle count from the RON is $CC_{RON} = \sum_{m=1}^{N_{RO}} CC_m$. Then, the data from the RO_i (CC_i) and RO_j $(j \neq i)$ (CC_j) are selected to calculate $x_i = (CC_{RON} - CC_i)/C_i$ and $y_j = (CC_{RON} - CC_i)/CC_j$. Finally, (x_i, y_j, I) from all the Trojan-free ICs would be used to generate the power signature, PS_{ij}. There will be $N_{RO} \times (N_{RO} - 1)$ unique power signatures in total. The same process will be applied to the CUT (shown in Fig. 6.8b). If the CUT lies within the signature, it may be assumed that the circuit is Trojan-free. Otherwise, if one of the $N_{RO} \times (N_{RO} - 1)$ signatures does not match the Trojan-free signature, it will be treated as a suspicious part, i.e., Trojan-inserted.

a

Power Signature Generation

Collect data from N_{TF} Trojan-free ICs with N_{RO} ring oscillators
for $(i = 1, i <= N_{RO}, i++)$ { //select i^{th} RO
  for $(j = 1, j <= N_{RO}, j++)$ $(j \neq i)$ { //select j^{th} RO
    for $(k = 1, k <= N_{TF}, k++)$ { //select k^{th} Trojan-free IC
      $x_{ki}=(\sum_{m=1}^{N_{RO}} CC_{km} - CC_{ki})/CC_{ki}$;
      $y_{kj}=(\sum_{m=1}^{N_{RO}} CC_{km} - CC_{ki})/CC_{kj}$;
      plot(x_{ki}, y_{kj}, I_K);
    }
    The power signature, PS_{ij}, is created from all N_{TF} ICs.
  } //x_{ki} is named as the first vector
} //y_{kj} is named as the second vector

Note: CC_{km}, CC_{ki}, CC_{kj}: Oscillation cycle count of RO_m, RO_i, RO_j in k^{th} IC
I_k: The dynamic current of kth IC

b

Authentication

For each IC under authentication:
Collect data from N_{RO} ring oscillators ($CC_{RON} = \sum_{i=1}^{N_{RO}} CC_i$)
for $(i = 1, i <= N_{RO}, i++)$ {
  for $(j = 1, j <= N_{RO}, j++)$ $(j \neq i)$ {
    $x = (CC_{RON} - CC_i)/CC_i$;
    $y = (CC_{RON} - CC_i)/CC_j$;
    plot(x,y,I);
    if $((x,y)$ is outside of the power signature PS_{ij}
      {The IC is Trojan-inserted; Break; }
    else go on;
  }
}

Note: CC_i, CC_j: Oscillation cycle count of RO_i, RO_j
I: Dynamic current of the under test IC

Fig. 6.8 Advanced outlier analysis procedure

6.5 Simulation Results and FPGA Implementation Analysis

The proposed approach is implemented on a small s9234 benchmark using Synopsys 90nm technology and a larger circuit, the AES benchmark, on Xilinx Spartan-6 FPGAs (45nm technology). For IC simulation, the s9234 benchmark was designed with two vertical power straps and 35 rows, with $N_{RO} = 15$ ring oscillators constituting the on-chip structure. Twenty Trojans (T_1 to T_{20}) with different sizes, gates types, and physical distributions were inserted into s9234. Table 6.1 shows these twenty Trojans where FF represents a flip-flop, $Cen.$ indicates that the Trojan is centrally located, and $Dis.$ indicates that the Trojan is physically distributed (shown in Fig. 6.9). Ten combinational Trojans (T_1–T_{10}) tap internal signals working as comparators, and the sequential Trojans (T_{11}–T_{20}) act as shift registers. None of

Table 6.1 Twenty Trojans inserted in s9234 circuit

	Combinational Trojans									
	T_1	T_2	T_3	T_4	T_5	T_6	T_7	T_8	T_9	T_{10}
Sizes	2 gates	3 gates	4 gates	5 gates	7 gates	8 gates	10 gates	12 gates	16 gates	20 gates
Area overhead (%)	0.09	0.16	0.2	0.25	0.37	0.43	0.5	0.66	0.82	0.92
Distribution	Cen.	Cen.	Dis.	Dis.	Cen.	Dis.	Dis.	Cen.	Dis.	Dis.

	Sequential Trojans									
	T_{11}	T_{12}	T_{13}	T_{14}	T_{15}	T_{16}	T_{17}	T_{18}	T_{19}	T_{20}
Sizes	2 FFs	3 FFs	4 FFs	5 FFs	6 FFs	7 FFs	8 FFs	10 FFs	12 FFs	16 FFs
Area overhead (%)	0.41	0.62	0.81	0.98	1.18	1.4	1.61	1.83	2	2.21
Distribution	Cen.	Cen.	Dis.	Dis.	Cen.	Dis.	Dis.	Dis.	Dis.	Dis.

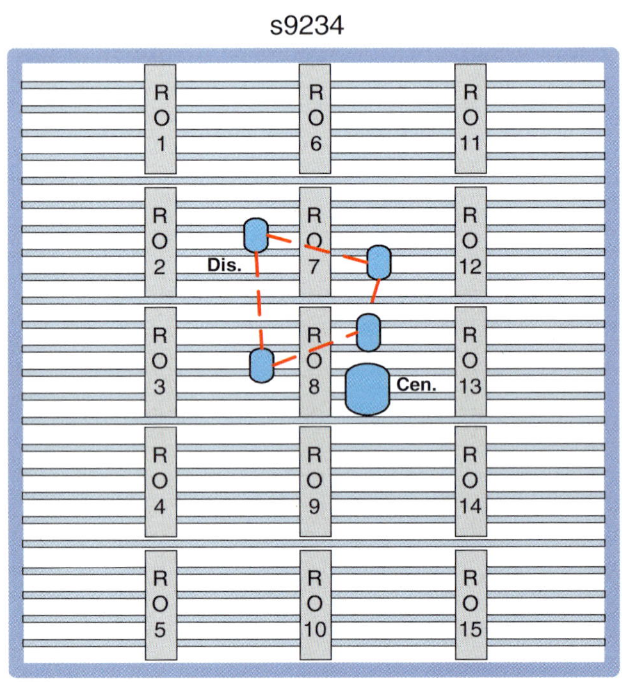

Fig. 6.9 s9234 with 15 ROs and 20 Trojans. One Trojan at a time is inserted into the circuit

these Trojans were detected by a test suite made up of 80,000 random functional patterns and 206 structural patterns (created by ATPG tools) for detecting stuck-at and transition delay faults. StarRC was used to extract parasitic parameters from the layout of benchmarks and to generate SPICE files. A Monte Carlo simulation (performed with Synopsys Nanosim) was used to emulate the effects of process variations that impact the frequencies of the ring oscillators and the dynamic current. The simulation temperature was 25°C with ±5°C variations. For hardware

Table 6.2 The oscillation cycle count of some of the ring oscillators and the circuit dynamic current in the presence of hardware Trojans without process variations

		T_1			T_3			T_6			T_{10}		
		TF	TI	ΔT	TF	TI	ΔT	TF	TI	ΔT	TF	TI	ΔT
Average dynamic current (μA)		29.8	29.84	0.04	25.56	25.65	0.09	24.94	25.08	0.14	24.94	26.1	1.16
RO (CC)	RO8	2790	2,787	−3	3,396	3,392	−5	3,064	3,054	−10	3,064	3,024	−40
	RO7	3,021	3,021	0	3,528	3,528	−2	3,008	3,005	−3	3,008	2,998	−10
	RO1	2,952	2,952	0	3,377	3,377	0	2,985	2,984	−1	2,985	2,982	−3
	RO15	3,103	3,103	0	3,406	3,406	0	2,803	2,803	0	2,803	2,801	−2

		T_{11}			T_{16}			T_{20}		
		TF	TI	ΔT	TF	TI	ΔT	TF	TI	ΔT
Average dynamic current (μA)		27.48	27.6	0.12	23.14	23.77	0.63	26.85	29.05	2.25
RO (CC)	RO8	3,150	3,141	−9	3,120	3,084	−36	3,031	2,972	−59
	RO7	3,117	3,117	−0	3,158	3,150	−8	2,925	2,914	−11
	RO1	3,042	3,042	0	3,198	3,198	0	2,980	2,977	−3
	RO15	3,132	3,132	0	3,210	3,210	0	3,012	3,011	−1

validation, eight Trojans (T_{21}–T_{28}) with different gates and distributions were inserted into an AES benchmark. Trojan-inserted and Trojan-free versions of the AES benchmark were both implemented on multiple FPGAs at room temperature; multiple FPGAs were used to analyze the effects of both inter-die and intra-die process variations.

6.5.1 Effectiveness Demonstration

Trojan Size and Distribution Analysis: Using a simulation without variations, the detailed cycle count and dynamic current results of T_1–T_3, T_6, T_7, and T_{12} with four ring oscillators (RO1, RO7, RO8, and RO15) during a 1,000-clock cycle LFSR test are shown in Table 6.2. Since the IC's dynamic current varies with the test pattern applied, the waveform of the dynamic current is recorded during the simulation. The average dynamic current in the measurement time window is shown in the table as well. In Table 6.2, TF indicates that the data in this column was collected from Trojan-free ICs while TI denotes data from Trojan-inserted ICs. ΔT represents the difference between the Trojan-inserted ICs and the Trojan-free ICs. As can be seen from Table 6.2, the Trojans consume extra power, increase the dynamic current, and decrease the cycle count of the ring oscillators.

Table 6.2 shows that T_1, T_3, and T_{11} have a larger impact on the oscillation frequency of RO8 than the other ring oscillators. Similarly, for T_6, T_{10}, T_{16} and T_{20}, there is a larger impact on RO8 and RO7 than on RO1 and RO15. This phenomenon is explained by the power supply voltage's dependence on the voltage division coefficient which is partially determined by the distance (resistive path) between two gates; a smaller distance implies a greater Trojan impact on ring oscillators. The remaining Trojans not shown in Table 6.2 exhibit similar behavior upon ring oscillators. However, the total dynamic current does not vary with the distributions of Trojans.

As Table 6.2 shows, in these seven Trojans, the oscillation cycle count difference ΔCC_{ft} of RO8 increased with Trojan size from −3 (for T_1) to −59 (for T_{20}). This occurred due to the greater power supply noise imparted from the Trojan with more gates. The dynamic current difference between a Trojan-free IC and a Trojan-inserted IC varies from 0.04 to 2.25 μA. Larger Trojans consume more power. Similar results can be observed for the Trojans not shown in the table. In general, larger Trojans have a greater impact on the power supply network, and consequently have a greater impact on the ring oscillators and dynamic current measurements.

Process Variations Analysis: Random process variations, consisting of 3σ = 10% voltage threshold (5% inter-die and 5% intra-die), 3σ = 3% oxide thickness (2% inter-die and 1% intra-die), and 3σ = 10% channel length (5% inter-die and 5% intra-die) are used in the following simulations to analyze their impact on the method. 200 Trojan-free ICs and 100 Trojan-inserted ICs for each Trojan are generated by Monte Carlo simulations. The statistical data analysis flow was used to process the data collected from these ICs. T_{10}, composed of 20 combinational gates, is used to show the detailed results of the data analysis flow.

A simple outlier analysis is first applied to distinguish the effects of Trojans and process variations. Histograms obtained from RO1, RO7, RO8, and RO15 in Fig. 6.10 show the distribution of oscillation cycle counts in the presence of process variations with T_{10}. Figure 6.10a displays a histogram of the cycle counts reported by RO8 with the inserted-Trojan, and Fig. 6.10b shows the same result without (w/o) the Trojan. The distributions of the two sets of oscillation cycle counts are plotted in Fig. 6.10c. The remaining figures (Fig. 6.10d–l) show the data distributions collected from RO7, RO1, and RO15, respectively. There is no significant change in RO7, RO1, and RO15. However, due to the presence of T_{10}, which is proximal to RO8, the RO8's distribution shifts leftward considerably. For RO8, the oscillation cycle range is 2756–3090 in Trojan-free ICs. 3 ICs out of the 100 ICs under authentication fell outside of the range, which are identified as containing a Trojan.

PCA is performed to analyze the data for the remaining 97 ICs. Figure 6.11a shows the power signature comparison using PCA with N_{RO} ring oscillators and the dynamic current for Trojan detection. The convex is drawn from the first three principal components with 200 Trojan-free ICs. The asterisks denote data obtained from ICs with Trojans that are shown to be separate from the convex hull. Thus, with the RON architecture and statistical analysis, T_{10} can be detected with 100% accuracy. However, with limited statistical analysis, or if the RON is subjected to the increasingly large variations of nano-scale technologies, *smaller Trojans* may not necessarily be detected with such accuracy, which was the case for T_1 to T_8 and T_{11} to T_{17}.

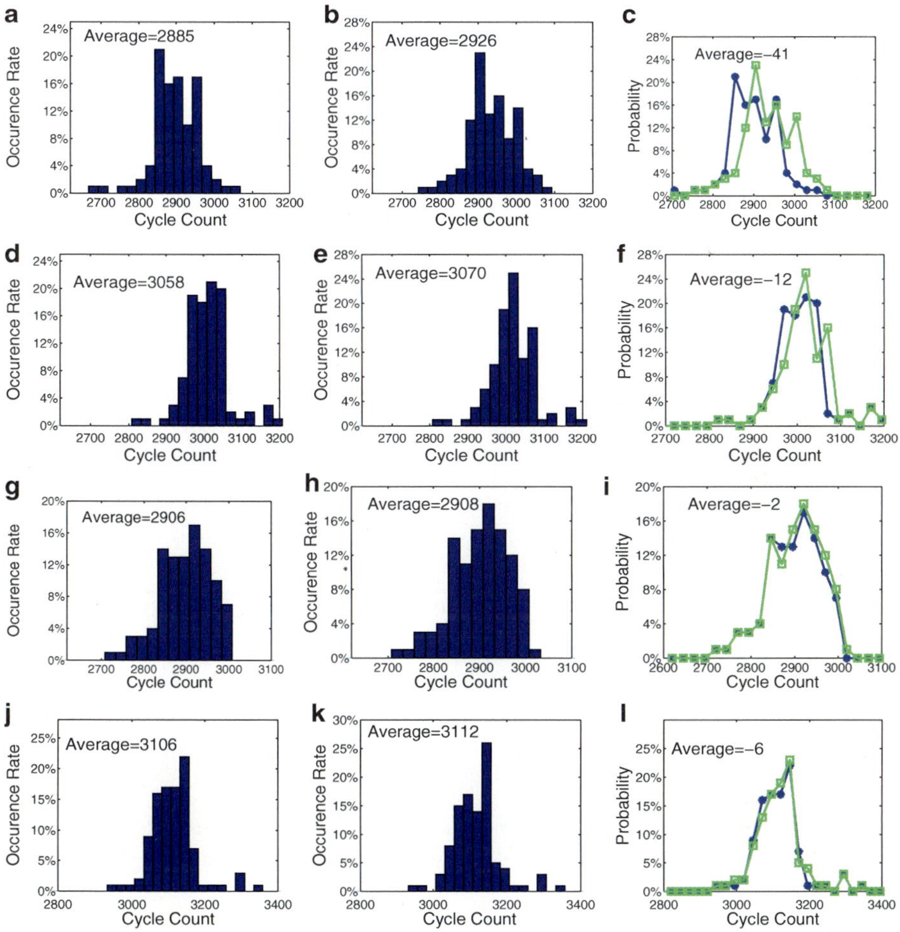

Fig. 6.10 Oscillation cycle distribution of RON with Monte Carlo simulations when T_{10} is inserted in s9234. (**a**) RO8 with Trojan; (**b**) RO8 w/o Trojan; (**c**) Cycle count distribution of RO8; (**d**) RO7 with Trojan; (**e**) RO7 w/o Trojan; (**f**) Cycle count distribution of RO7; (**g**) RO1 with Trojan; (**h**) RO1 w/o Trojan; (**i**) Cycle count distribution of RO1; (**j**) RO15 with Trojan; (**k**) RO15 w/o Trojan; (**l**) Cycle count distribution of RO15

The advanced outlier analysis shown in Fig. 6.8 is also used to identify Trojan-inserted ICs. There are a total of $15 \times 14 = 210$ power signatures generated by the Trojan-free ICs. Some power signatures could identify more Trojan-inserted ICs than others. In the following advanced outlier analysis results, only the power signature that can detect the most Trojan-inserted ICs is shown. Figure 6.11b shows the advanced outlier analysis results with Trojan T_{10}. The ring oscillator that was selected as x in Fig. 6.8 is defined as the first vector and y as the second vector. The

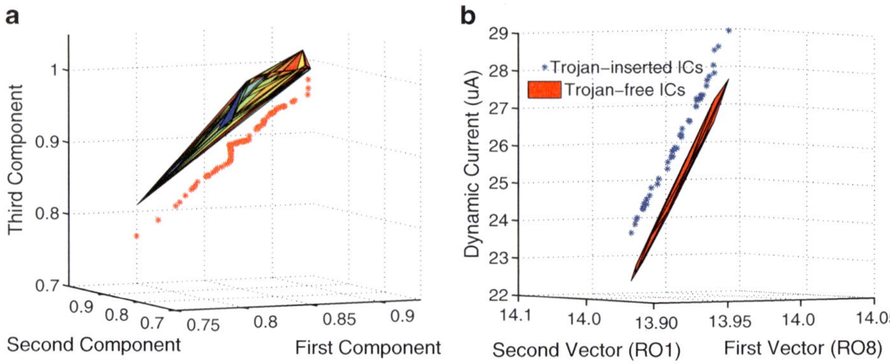

Fig. 6.11 Power signature for Trojan-free ICs and Trojan-inserted ICs with T_{10} using (**a**) PCA and (**b**) advanced outlier analysis

Table 6.3 Trojan detection rates with process variations

	Combinational Trojans									
	T_1	T_2	T_3	T_4	T_5	T_6	T_7	T_8	T_9	T_{10}
Trojan detection rate (%)	75	80	86	100	100	100	100	100	100	100
	Sequential Trojans									
	T_{11}	T_{12}	T_{13}	T_{14}	T_{15}	T_{16}	T_{17}	T_{18}	T_{19}	T_{20}
Trojan detection rate (%)	100	100	100	100	100	100	100	100	100	100

blue dots represent Trojan-free ICs and the red asterisks denote Trojan-inserted ICs. As can be seen in the figure, all of the Trojan-inserted ICs are outside of the Trojan-free signature. Thus, the detection rate with T_{10} using advanced outlier analysis is 100%.

Similarly, the remaining 19 Trojans with 200 Trojan-free ICs and 100 Trojan-inserted ICs are also simulated and the data analysis flow is applied for every Trojan. The Trojan-inserted ICs with T_1, T_4, T_{11}, and T_{20} are selected to present detailed results using advanced outlier analysis, shown in Fig. 6.12a–d. The detection rates of Trojans T_{11} and T_{20} shown in Fig. 6.12 are 100% with only one signature. For T_4, 98% of the Trojan-inserted ICs are detected using one signature, shown in Fig. 6.12b. When all 210 power signatures are used, the detection rate for Trojan T_4 is 100%. Complete results for all Trojans using all of the power signatures are shown in Table 6.3. From Table 6.3 and Fig. 6.12, it can be concluded that for Trojan T_4–T_{20}, the detection rates are all 100%. The power signatures of the Trojan-free ICs are completely separate from the Trojan-inserted ICs. However, the Trojan-inserted ICs with T_4 are close to the Trojan-free ICs. For the very small Trojans T_1–T_3, the detection rates are less than 100% because of their diminished impact on the power supply lines.

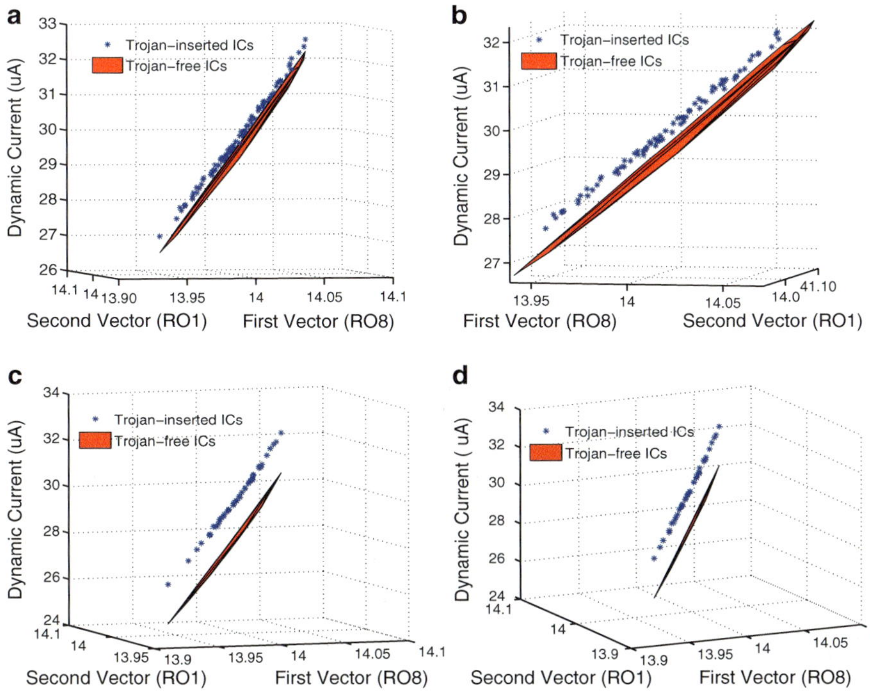

Fig. 6.12 Signatures with outlier data analysis from IC simulation for Trojan (**a**) T1, (**b**) T4, (**c**) T11 and (**d**) T20

6.5.2 *Sensitivity Analysis*

Ring Oscillator Number Analysis: Trojans T_1, T_2, and T_3 were chosen for ring oscillator number analysis since their detection rates are less than 100% with N_{RO}=15 ring oscillators. RONs with N_{RO}=10, 20, and 25 ring oscillators were implemented with Monte Carlo simulation. The location of the inserted Trojans is fixed throughout this analysis. For RONs with different quantities of ring oscillators, the layout is similar to Fig. 6.9. The three columns of ring oscillators were replaced by 2, 4, and 5 columns of ring oscillators.

Figure 6.13 shows Trojan detection rates using advanced outlier analysis with a different number of ring oscillators in RON for Trojans T_1, T_2, and T_3. With 10 ring oscillators, the detection rates for T_1, T_2, and T_3 are 40%, 48%, and 53%, respectively. With 25 ring oscillators, the detection rates increase to 95%, 100%, and 100%. These results imply that increasing the number of ring oscillators in the circuit improves detection rates. This is because a Trojan will likely be closer to a ring oscillator (or perhaps several) with more of them embedded in a design.

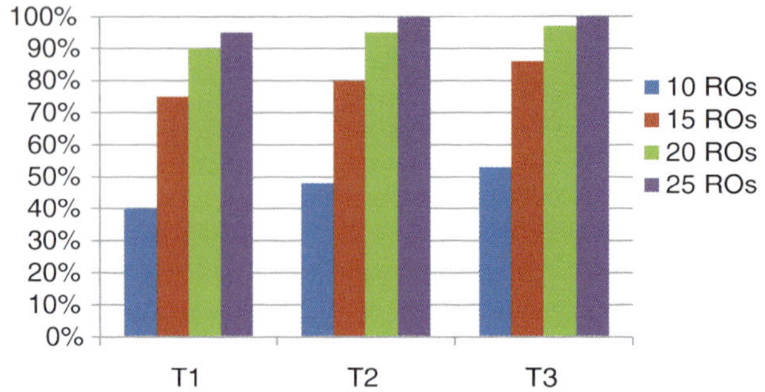

Fig. 6.13 Ring oscillator number (N_{RO}) analysis with Trojans T_1, T_2, and T_3

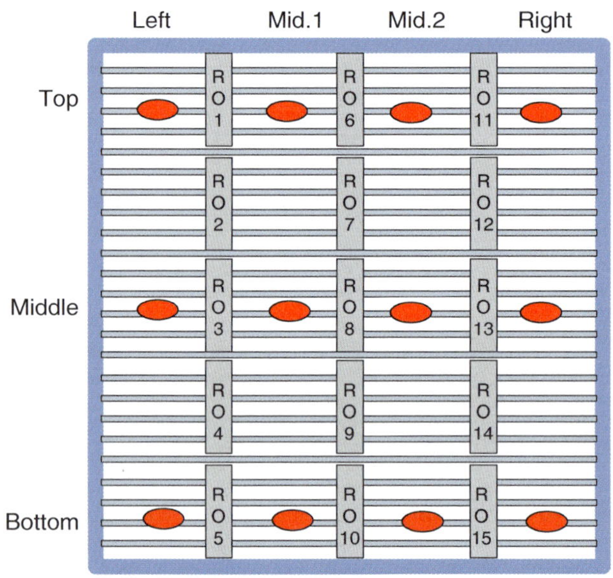

Fig. 6.14 Placing T_2 at different location in the s9234 circuit

When the number of ring oscillators in the RON is increased, the power consumption will remain unchanged while the circuit is under normal operation; the RON is only on for a short time during testing and remains off during functional operation. The area overhead would increase slowly with the number of ring oscillators.

For the simulation, the area overheads are 2.5%, 3.75%, 5.0%, and 6.25% with 10, 15, 20, and 25 ring oscillators in the RON inserted in s9234. However, the increase in area overhead is small in comparison to the increase in Trojan detection rates. Thus, the RON structure may be adjusted to meet desired area overhead and detection resolution values.

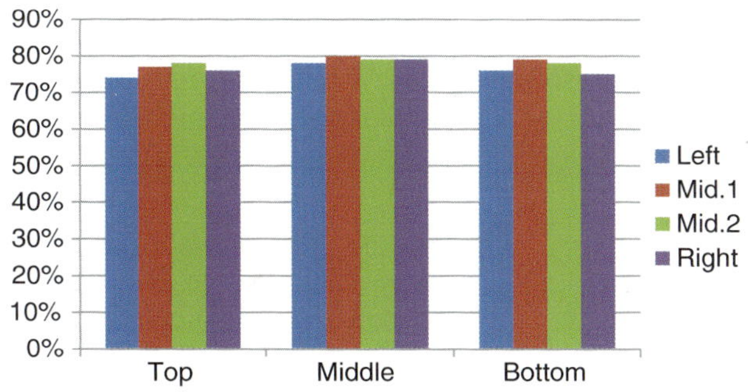

Fig. 6.15 Trojan location analysis with T_2

Trojan Location Analysis: In order to verify the impact of a Trojan's location on its detection rate, Trojan T_2 was placed in twelve locations (shown in Fig. 6.14). For each location, 200 Trojan-free ICs and 100 Trojan-inserted ICs were generated by a Monte Carlo simulation. A RON of 15 ROs was embedded into the benchmark. The detection rates using advanced outlier analysis are shown in Fig. 6.15. As the figure shows, when Trojan T_2 was placed around boundary corners, fewer Trojan-inserted ICs would be detected than if it was placed centrally. This occurs because the Trojan is closer to a greater number of ring oscillators when placed towards the center. However, the Trojan detection rate varies by less than 8% for the twelve locations. This can likely be alleviated with greater design coverage; placing ring oscillators in columns adjacent to the outermost edges of the IC will limit the maximum distance between a Trojan and an RO.

Pattern Analysis: Since different inputs could cause different switching activities within an IC, the pattern generated by the LFSR during testing can impact Trojan detection resolution in two ways: (1) Trojan switching activity (and thus the Trojan contribution to changes in dynamic power) depends on circuit inputs and thus the pattern selected, and (2) the total switching activity in the circuit may be altered by the patterns. Increased switching among Trojan gates implies a greater Trojan contribution to side-channel information. Decreased total switching in the circuit under authentication implies reduced background noise and a greater chance that Trojan activity will not be obfuscated. It is crucial to note that Trojan switching activity does not refer to the event of actually activating a Trojan to launch its malicious function, but rather refers to any amount of switching in the gates which comprise the Trojan. For example, for Trojan T_3, which is composed of four gates, if only one gate transitions, there is switching activity in the Trojan, whether or not the Trojan was completely activated. The LFSR was simulated to verify the impact of pattern selection on the combined ring oscillator network and the dynamic current method. Different seeds are used in the LFSR to generate different patterns; 1,000 patterns are generated by one seed.

Figure 6.16 shows the detection rates with four different seeds ($S1$, $S2$, $S3$, and $S4$) in the LFSR. The four different seeds were randomly generated by MATLAB.

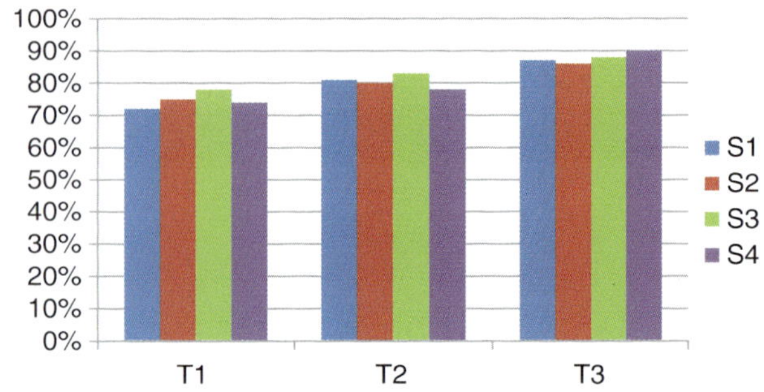

Fig. 6.16 Pattern analysis with Trojans T_1, T_2, and T_3

Trojans T_1, T_2, and T_3 were selected to show the results. All these Trojans were fixed at locations shown in Fig. 6.9. As seen in Fig. 6.16, the Trojan detection method gives different detection rates using different random patterns. Generally, the detection rate will be higher if the Trojan switching activity is greater. However, the Trojan detection rated does not vary significantly with random patterns. If special patterns are generated, such as ones that could cause more switching at the nets that rarely activate in the design, the Trojan detection method would be more effective.

6.5.3 Experimental Results from Spartan-6 FPGA

Xilinx Spartan-6 FPGA boards (shown in Fig. 6.17a) were used for the hardware validation of the proposed method and 24 ring oscillators were inserted into an AES benchmark circuit (shown in Fig. 6.17b). An Atmel Atmega328P microcontroller is connected to the FPGA to facilitate the collection of ring oscillator cycle count data from the counter. Transient current waveforms (shown in Fig. 6.18a) are collected using Digilent Adept software [8]. 28 Trojan-inserted FPGAs and 60 Trojan-free FPGAs were used to verify the impact of process variations. Several measurements were done for each ring oscillator in each FPGA in order to eliminate measurement noise, and the average oscillation count was used to perform data analysis. The Trojans implemented in the following analysis are composed of arbitrary combinational gates of varied sizes. The malicious function to be carried out by the Trojans will not be important since this analysis is intended to demonstrate the ability of the method to detect arbitrarily added yet difficult to detect malicious gates.

Eight different Trojans T_{21}–T_{28} with different sizes were inserted into the AES benchmark. As seen in Table 6.4, some Trojans are extremely small and switch rarely during functional operation. For example, the switching probability of Trojan T_{21} is 0.0016%. These Trojans were found at location L_3 (shown in Fig. 6.17b). The area overhead and detection rates of these Trojans are shown in Table 6.4. T_{26} was used to show the detailed results of the advanced outlier analysis in Fig. 6.18b. As can be seen from the figure, the Trojan detection rate for T_{26} is 100%. With all the Trojan detection rates (shown in Table 6.4), it can be concluded that most of

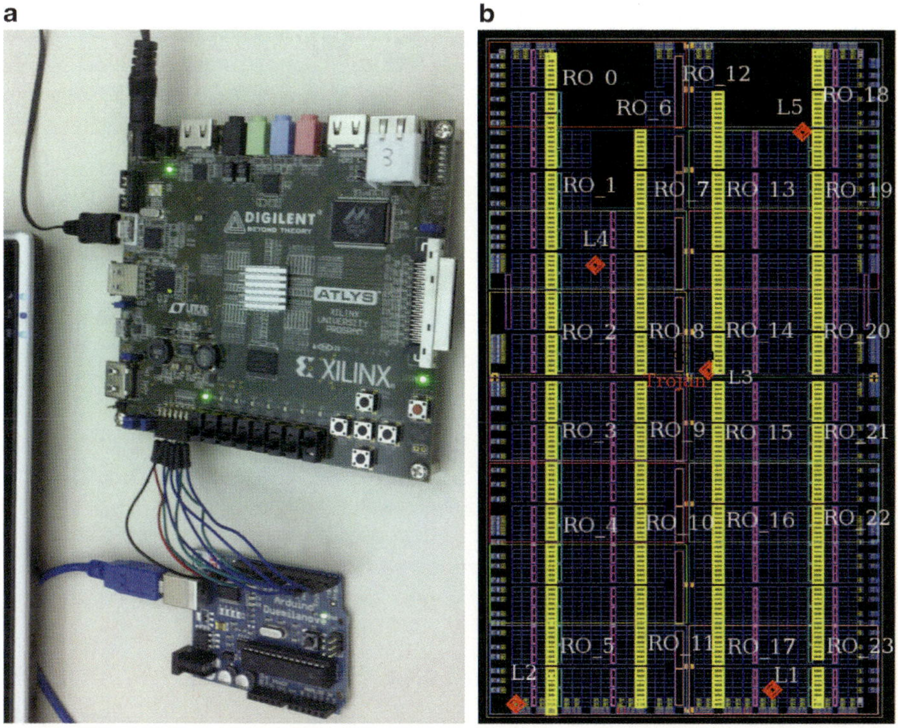

Fig. 6.17 (**a**) Xilinx Spartan-6 FPGA board (45nm technology) and (**b**) AES layout after placement

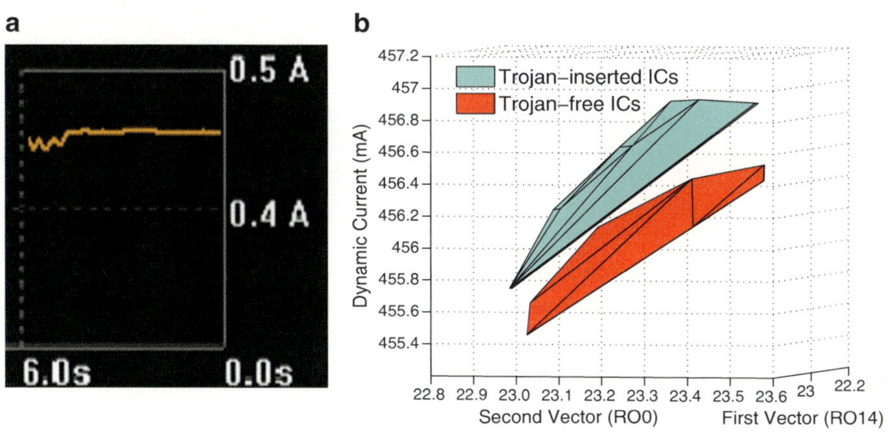

T_{26} T_{26}

Fig. 6.18 (**a**) Transient current waveform and (**b**) Outlier analysis results with Trojan T_{26} from FPGA implementation

Table 6.4 Trojans inserted in FPGAs and their detection rate when $N_{RO} = 24$

	T_{21}	T_{22}	T_{23}	T_{24}	T_{25}	T_{26}	T_{27}	T_{28}
Area overhead (%)	0.0016	0.012	0.025	0.05	0.08	0.1	0.15	0.2
Trojan detection rate (%)	73	86	100	100	100	100	100	100

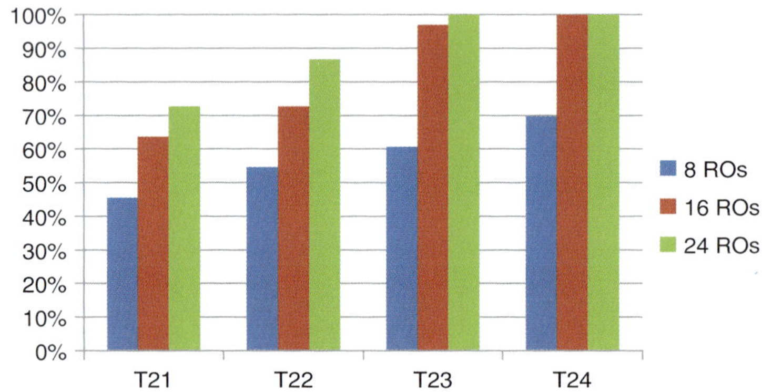

Fig. 6.19 Ring oscillator number analysis with Trojans T_{21}, T_{22}, T_{23} and T_{24} in FPGAs

Trojans were detected with a 100% detection rate. However, for very small Trojans, the detection rates were lower.

The impact of the number of ring oscillators on detection rates was analyzed on Xilinx Spartan-6 FPGAs, in addition to the simulation results presented earlier. Here, the number of oscillators in the network is varied and Trojans of diverse sizes are inserted into the circuit. The Trojans are placed in the same location, and the same LFSR seed is applied to each part of this experiment. RONs, composed of 8, 16, and 24 ring oscillators, were implemented in the AES benchmark circuit. Figure 6.17b shows the RON with 24 ring oscillators and RONs with 8 ring oscillators and 16 ring oscillators similarly implemented. With 60 Trojan-free FPGAs and 28 Trojan-inserted FPGAs, Fig. 6.19 shows detection rates with different RONs for Trojans T_{21}, T_{22}, T_{23}, and T_{24}. The figure shows that the number of ring oscillators in the RON plays a considerable role in the effectiveness of the method. For T_{23} and T_{24}, a detection rate of 100% is achieved by increasing the size of the network from 8 to 24 ring oscillators.

Also, a significant improvement is achieved by increasing the number of ROs from 8 to 16, but a smaller improvement is seen when the number of ROs is increased from 16 to 24. This suggests that detection resolution is not linear with the number of ring oscillators in RON.

To analyze the sensitivity of the method to the location of Trojans, T_{22} was placed in different locations, from L_1 to L_5. Figure 6.20 shows results using the data analysis flow. The detection rate vacillates between 88.3% and 79.3% with changes in the Trojan's location. When the Trojan was placed in locations L_4 and L_3, the detection rate was relatively higher, since it impacted more ring oscillators.

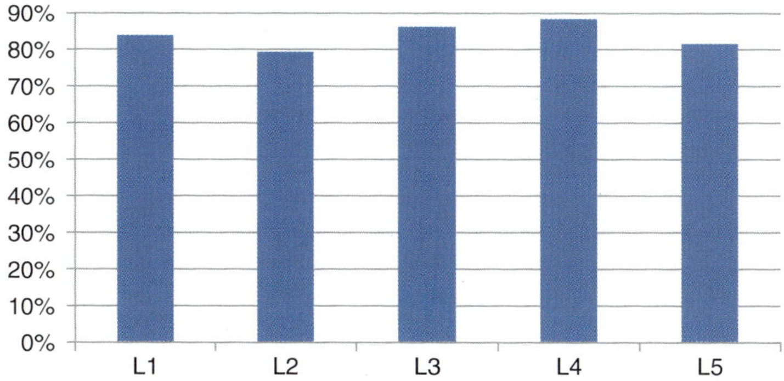

Fig. 6.20 Trojan location analysis with Trojans T_{22} in FPGAs

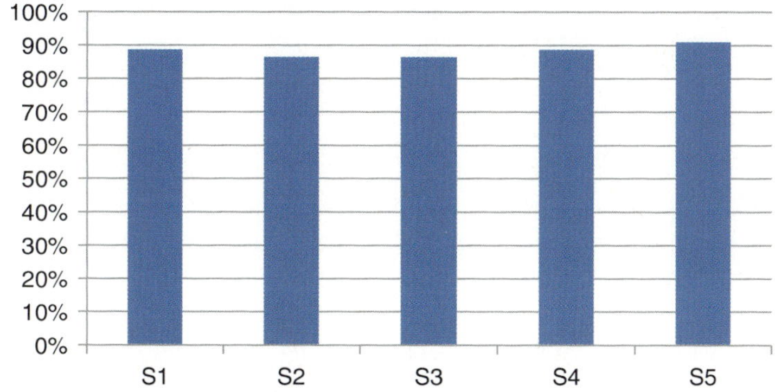

Fig. 6.21 Patterns analysis with Trojans T_{22} in FPGAs

When the Trojan was located in L_2, at a corner of the FPGA, the Trojan detection rate is at its lowest.

To analyze the impact of these patterns, Trojan T_{22} was located in L_3. Six randomly selected seeds were applied to the LFSR. The ring oscillator cycle counts and transient current waveforms were collected and analyzed. Figure 6.21 shows the data analysis results. The figure shows that random patterns do not have a significant impact on the Trojan detection rate. However, if a designer were to intelligently select a set of patterns that control background noise and net coverage, additional improvements in detection resolution are possible.

6.6 ASIC Evaluation

6.6.1 Test Chip Design

In order to analyze the effectiveness of the RON structure, 40 test chips were designed and fabricated using IBM 90nm technology through MOSIS. All chips

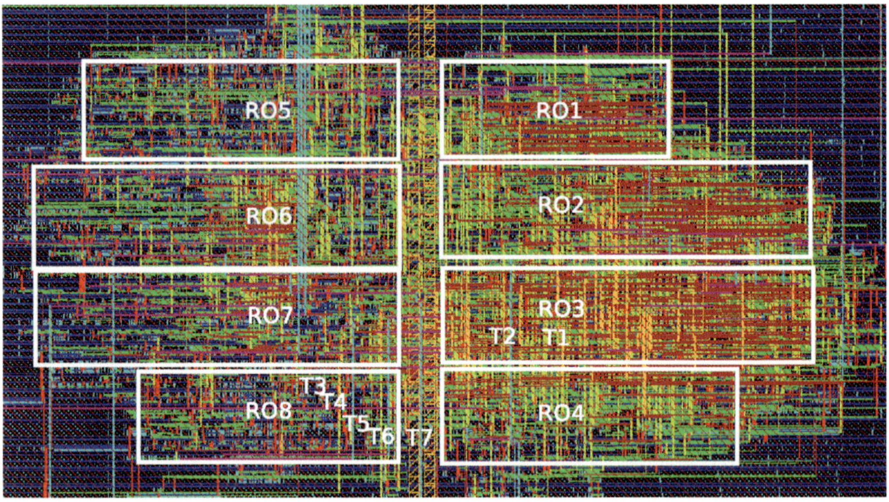

Fig. 6.22 Layout for the test chip design

were fabricated on the same wafer. The RON architecture is inserted into the ISCAS s9234 benchmark, which represents the design to be protected in the test chip.

Figure 6.22 shows the layout of the test chips with the RON structure composed of $N_{ro} = 8$ $n = 61$-stage ROs (RO_j, where $1 \leq j \leq 8$) with one NAND gate and 60 inverters, each distributed across the chip. It is important to note that the areas labeled RO_1 to RO_8 show the broad area in which that RO is confined rather than the total area occupied by that RO. Ring oscillator stages are placed in each standard cell row in an intentionally loosely distributed fashion that improves its coverage of the power distribution network. Therefore, these areas are also occupied by background circuit and control structure components and the area overhead of the oscillators is substantially lower than the labeled areas. The approximate locations of the seven Trojan stages (T_i where $1 \leq i \leq 7$) are labeled as well. The number of RO stages was selected so that the maximum observed frequency would not exceed the 400 MHz operating frequency of the 90 nm counters used in this design. The distance between the two adjacent RO components is limited to 10 times the width of the flip-flops. Given this design rule and the area of the chip, 8 ROs were used.

The feedback polynomial of the LFSR used in the test chip is

$$X^7 + X^3 + 1 \tag{6.10}$$

To conserve area, this design uses an LFSR with only 8-bits to generate patterns for the 36 input s9234 benchmark. A broadcasting technique is used to assign this 8-bit output to the 36 inputs. An 8-bit decoder and 8-bit multiplexer are used for RO selection. A 16-bit counter is used to measure the number of oscillations observed in the test duration, which is controlled by a timer. In this design, the test duration of 500 clock cycles was selected based on the technology node and test area overhead.

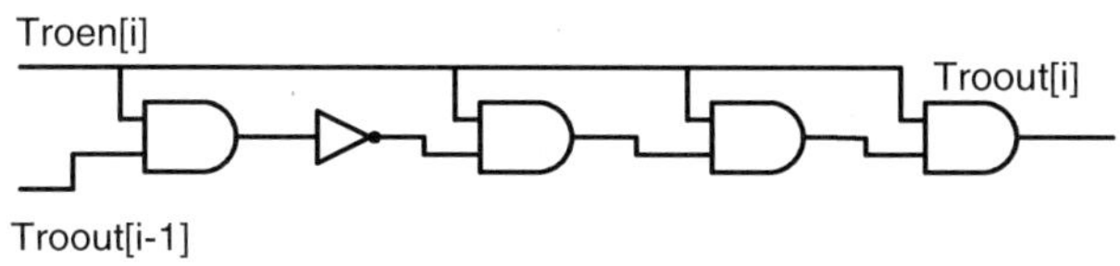

Fig. 6.23 Design of a hardware Trojan stage T_i

Table 6.5 An estimation of the area occupied by s9234 in terms of the number of transistors

Component	Quantity	Total transistors
D Flip-Flops	211	7,174
Inverters	3,570	7,140
Gates	2,027	8,108
Total	5,808	22,422

6.6.2 Hardware Trojan Design

Each IC contains seven combinational hardware Trojan designs that may be completely deactivated. Since this design is implemented in 90nm CMOS technology, the static power dissipation and thus the side-channel contribution is negligible when the Trojans are deactivated. By using a single-IC multiple-Trojan design we are able to not only carry out a more extensive set of Trojan impact tests, but we are also able to isolate the effect of process variations from the effect of inserted Trojans on RO characteristic frequencies. Further, since the static power is present in the Trojan-free case, it is neglected and the detection results provide a lower-bound.

The gate-level implementation of a Trojan stage is shown in Fig. 6.23 where $troout[i]$ is the output of the ith Trojan stage, $troout[i-1]$ is the output of the previous Trojan stage, and $troen[i]$ is the enable signal for the ith stage which also asserts all prior enable signals when enabled.

Trojan T_i contains i stages consisting of $i \times (4AND + 1INV)$ gates where each stage $i-1$ is also enabled if stage i is enabled. The first Trojan, T_1 is driven by the 200 MHz clock signal at the location of signal $troout[0]$. Note that the Trojan T_i is not derived of the trigger-payload Trojan design used in [11–13]. Here, each Trojan gate transitions once per clock cycle; therefore, the partial activity of each of these Trojans is simply $5i$ partial activations per clock cycle. The average ratio of Trojan partial activation to background circuit activity is estimated in the fourth column of Table 6.6.

The s9234 benchmark consists of 211 D flip-flops, 3570 inverters, and 2027 other gates. The number of transistors used in the s9234 benchmark is estimated in Table 6.5 by assuming that each flip-flop consists of 8 NAND or NOR gates and 2 inverters. As previously mentioned, there are a total of seven Trojans (T_1–T_7) in this design. The area overhead of each Trojan is summarized in Table 6.6.

Table 6.6 An estimation of Trojan area overheads and noise

Trojan number	Transistors	Percent area (%)	Trojan to background circuit switching ratio (%)
T1	26	0.12	0.11
T2	52	0.23	0.22
T3	78	0.35	0.33
T4	104	0.47	0.45
T5	130	0.58	0.56
T6	156	0.70	0.67
T7	182	0.81	0.78

Fig. 6.24 A data collection setup including a Spartan 6 FPGA connected to a prototyping board through a serial connector

6.6.3 Experimental Setup

During data collection, the IC is mounted on and wired to a prototyping board that includes a high-density serial connector. The serial connector allows the prototyping board to interface with a Xilinx Spartan-6 FPGA on a Digilent Nexys 3 board. The FPGA is programmed to control the test sequence supplied to the IC and to transmit the outputs of the IC to a computer using an on-board USB-UART module. The complete setup is shown in Fig. 6.24.

The nominal supply voltage of the IC pins is 2.5V. This is converted internally to the nominal core voltage of 1.2V using a voltage divider. Since the s9234 benchmark circuit used in this design is small compared to a modern IC, in order to emulate the tight power design of a modern circuit, an external voltage divider is used to supply the IC with 1.875V and the core with 0.9V, which is greater than the 0.80V minimum core voltage. Reducing the power supply voltage will reduce the background circuit switching activity and improve Trojan detection rates. Therefore, it is desirable to reduce the supply voltage during measurement.

The FPGA includes a state machine which sequences through each ring oscillator, begins a data collection trial, selects each 4-bit window of the counter output for the current ring oscillator, and transmits each 4-bit window as a hex digit over the USB-UART connection. The process is repeated for 10 trials on each ring oscillator of each IC. The IC is supplied with 1.875V using a voltage divider and the board's 2.5V peripheral power supply over the serial connection along with a 200 MHz clock signal. Each trial lasts 500 clock cycles.

As shown in Fig. 6.22, each of the 40 ICs contains $N_T = 7$ pre-inserted hardware Trojan designs. During Trojan-free data collection, each hardware Trojan circuit is disabled, as is any Trojan not being analyzed. Since the designs are implemented with CMOS circuits, the static dissipation is negligibly low. Furthermore, since all Trojan measurements are compared to the Trojan-free results (which include static dissipation), the detection results provide a conservative lower bound.

6.6.4 Experimental Results and Analysis

The frequency of a single ring oscillator on a single IC was measured 10 times. The measurement noise is then calculated with

$$\frac{Max\{f_{Trial1}, \ldots, f_{Trial10}\} - Min\{f_{Trial1}, \ldots, f_{Trial10}\}}{0.1 \sum_{m=1}^{10} f_{Trialm}} \tag{6.11}$$

for a single IC and a single ring oscillator where f_{Trialm} is the mth repeated measurement of frequency for that RO. This is repeated for all ICs and all ROs and averaged, resulting in a measurement noise of 0.23%.

The impact of intra-die variation on an RO's frequencies was analyzed by comparing a single RO on an IC with other ROs on that same IC. For a single IC, intra-die variation is calculated with

$$\frac{Max\{f_{RO_1}, \ldots, f_{RO_8}\} - Min\{f_{RO_1}, \ldots, f_{RO_8}\}}{0.125 \sum_{j=1}^{8} f_{RO_j}} \tag{6.12}$$

where f_{ROj} is the frequency of the jth RO. This calculation is repeated for all ICs and averaged, resulting in a mean intra-die variation impact on frequency of 8.05%.

Of the 40 fabricated ICs, 38 functioned correctly and the remaining faulty ICs were omitted. The impact of inter-die variation on the frequency of a ring oscillator was determined by selecting a single RO and comparing the frequency of this RO across each IC. For a single RO, the inter-die variation is calculated with

$$\frac{Max\{f_{IC1}, \ldots, f_{IC38}\} - Min\{IC_1, \ldots, f_{IC38}\}}{(1/38) \sum_{k=1}^{38} f_{ICk}} \tag{6.13}$$

Table 6.7 A summary of validation data

Measurement noise	0.23%
Intra-die variation	8.05%
Inter-die variation	16.67%
Mean RO frequency	291 MHz

where f_{ICk} is the frequency of the individual RO of interest on the kth integrated circuit. This calculation is repeated for all ROs and averaged, resulting in a mean inter-die variation impact on frequency of 16.67%. The average RO frequency of all ROs on all ICs was 291 MHz. The maximum recorded frequency was 315 MHz, which was less than the 400 MHz frequency the counter was timing closed at. These results are summarized in Table 6.7.

Trojan Impact Analysis: The direct impact of hardware Trojan induced power supply noise on ring oscillator frequencies is analyzed by measuring the frequency of each RO on each IC for the Trojan-free case, as well as for each Trojan. The mean impact of a particular Trojan on a particular RO is then computed by comparing the frequency of that RO on a particular IC with the frequency of that RO on the same IC when the Trojan is disabled. The computation is thus

$$TROI_{ROj,Ti} = (1/38) \sum_{k=1}^{k=38} \frac{|RO_{j,k,Tfree} - RO_{j,k,Ti}|}{RO_{j,k,Tfree}} \times 100\% \quad (6.14)$$

where $TROI_{ROj,Ti}$ is the mean impact of the ith Trojan on the jth RO across all ICs compared to the Trojan-free case. $RO_{j,k,Tfree}$ is the Trojan-free frequency for the jth RO on the kth IC, and similarly, $RO_{j,k,Tj}$ is the frequency of the jth RO on the kth IC with the ith Trojan activated.

It is with this calculation that the value of the single-IC multiple-Trojan design is best demonstrated. By comparing measurements made with a Trojan enabled against measurements made on the same IC with the Trojan disabled, inter-die variation is eliminated from the analysis. Had separate ICs been fabricated with Trojans inserted and Trojans removed, only comparisons between different ICs would be possible, and the computation would include inter-die process variation. By restricting comparisons to the same RO, intra-die process variations are eliminated from the computation as well.

The results for Trojan impact are presented in Fig. 6.25. It is immediately clear that Trojans of greater area and those that more frequently partially activate induce a greater change in the frequencies of nearby ROs since they consume more power. The maximum induced change for the largest Trojans in this experiment is representative of one of the core issues in the IC trust problem. The Trojan induces at most a change of 2.5% to frequencies, yet as Table 6.7 reports, intra-die variation and inter-die variation induce far greater changes, suggesting these Trojans would be completely obfuscated in a test where these variations are not isolated. However, Trojan detection is still possible with this technique. The manner in which Trojan

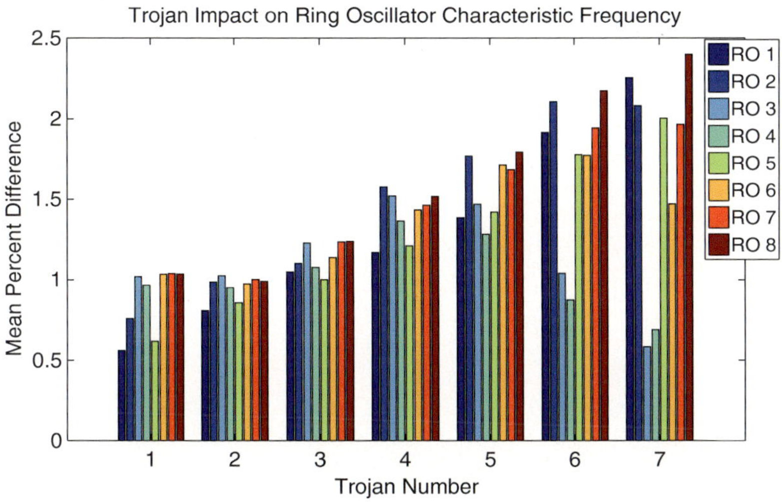

Fig. 6.25 The impact of inserted hardware Trojans on RO frequencies isolated from process variations

impact is distributed across ROs, including the decrease in impact on RO_3 and RO_4 for larger Trojans.

Spatial Locality Analysis: To analyze the effect of Trojan location, the ring oscillator that experiences the greatest Trojan impact calculated with (6.14) is determined for each IC with a particular Trojan. A histogram showing the frequency with which each ring oscillator was the most impacted on an IC is shown in Fig. 6.26. The location of Trojan gates relative to the gates of the ROs and the vertical power line is shown in Fig. 6.22

Notably, RO_8 is impacted most frequently of all Trojans since several of its gates are closest to the vertical power strap, thereby causing a portion of the overall power supply noise to affect this RO. For T_1 and T_2, a substantial portion of the Trojan impact is distributed on RO_2 and RO_3, since these Trojans are located close to these ROs and likely share power lines.

Since the majority of the gates in subsequent Trojans are closest to RO_8, more of the Trojan impact is distributed on this RO. Perhaps counter-intuitively, the distribution becomes more focused on a single RO as the Trojan expands in size. Had the Trojan expanded vertically and towards multiple ROs it is likely the distribution would have become less focused. However, for these Trojans that extend primarily horizontally, the increase in area and activity further increases the Trojan impact without expanding into other regions of the power network.

For T_7, the Trojan becomes less localized on RO_8 since T_7 is particularly close to the vertical power strap. For this reason, the Trojan impact is more evenly distributed across ROs since the vertical power strap supplies power to the entire circuit. Finally, the reduced impact on RO_3 and RO_4 for T_6 and T_7 shown in Fig. 6.25 is due to the

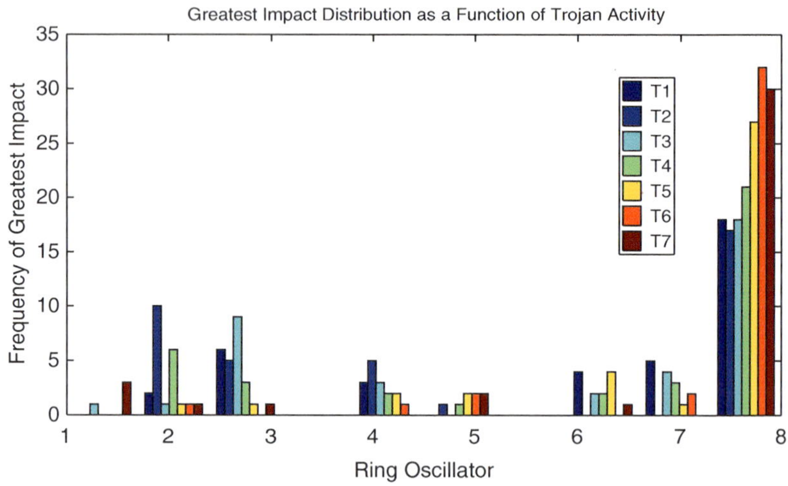

Fig. 6.26 The number of instances of each RO being most impacted by a Trojan

loosely distributed nature of these ROs away from the vertical power line and the placement of these Trojans close to the vertical power line.

IC Classification and False-Positive Analysis: In the previous section, it was shown that all Trojans used in this study impacted the RO frequencies substantially less than inter-die and intra-die process variations. However, using the principal component analysis (PCA) [4] based classification scheme presented below, it is still possible to detect these Trojans. In order to verify that this data is adequately represented in fewer than 8 principal components, the percent of the total variance in each PCA representation is computed by dividing the cumulative sum of the latent of the PCA representation by the total sum. The percent variance for each representation is shown in Table 6.8. The results imply that any representation of at least 2 components should adequately represent this data.

Table 6.8 The percent variation contained in a representation of h principal components

Components	Percent variation (%)
1	89.4
2	99.39
3	99.59
4	99.79
5	99.87
6	99.93
7	99.97
8	100

To succeed, a classification scheme must perform two functions: (1) it must correctly label Trojan-inserted circuits as tampered and (2) it must correctly label

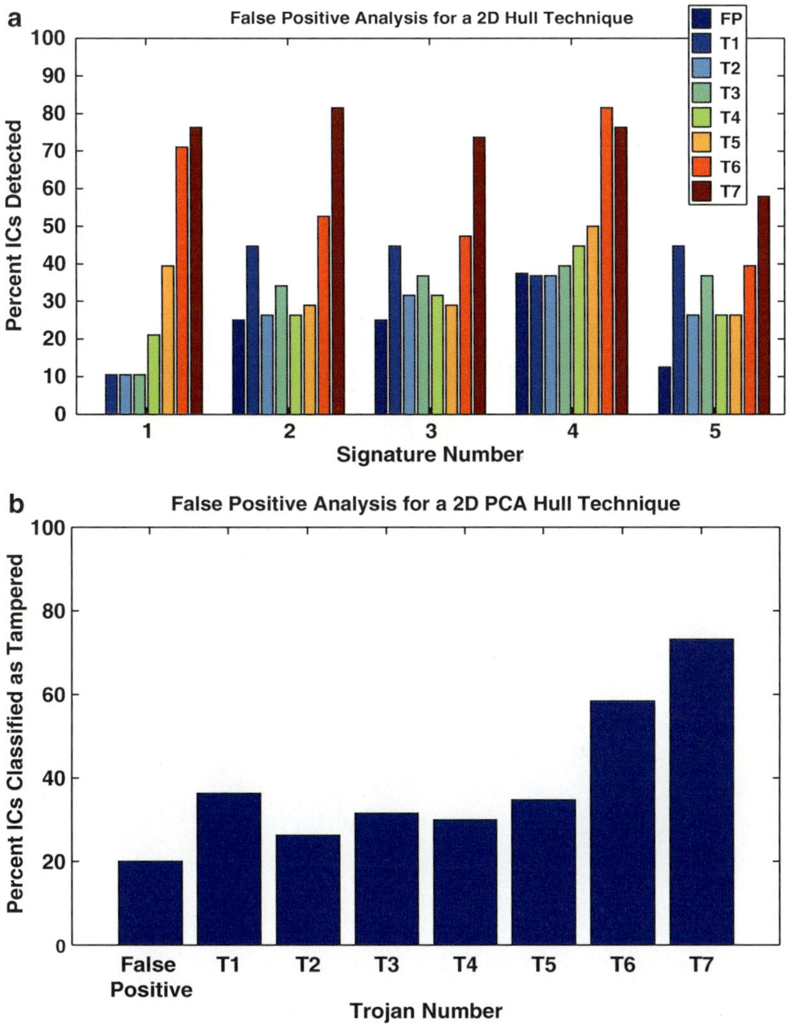

Fig. 6.27 Classification using the presented scheme and 2 dimensions. (**a**) All cases. (**b**) Mean rates

Trojan-free circuits as uncompromised. The steps for the presented classification scheme are:

1. Form a matrix from golden (Trojan-free) data in which each row is a verified Trojan-free IC and each column is a ring oscillator. Append a similar row containing the data from the chip under authentication (CUA) to the matrix.
2. Obtain a representation of this matrix using the first h principal components.
3. Render an h-dimensional convex hull [5] with all data, except that of the CUA.

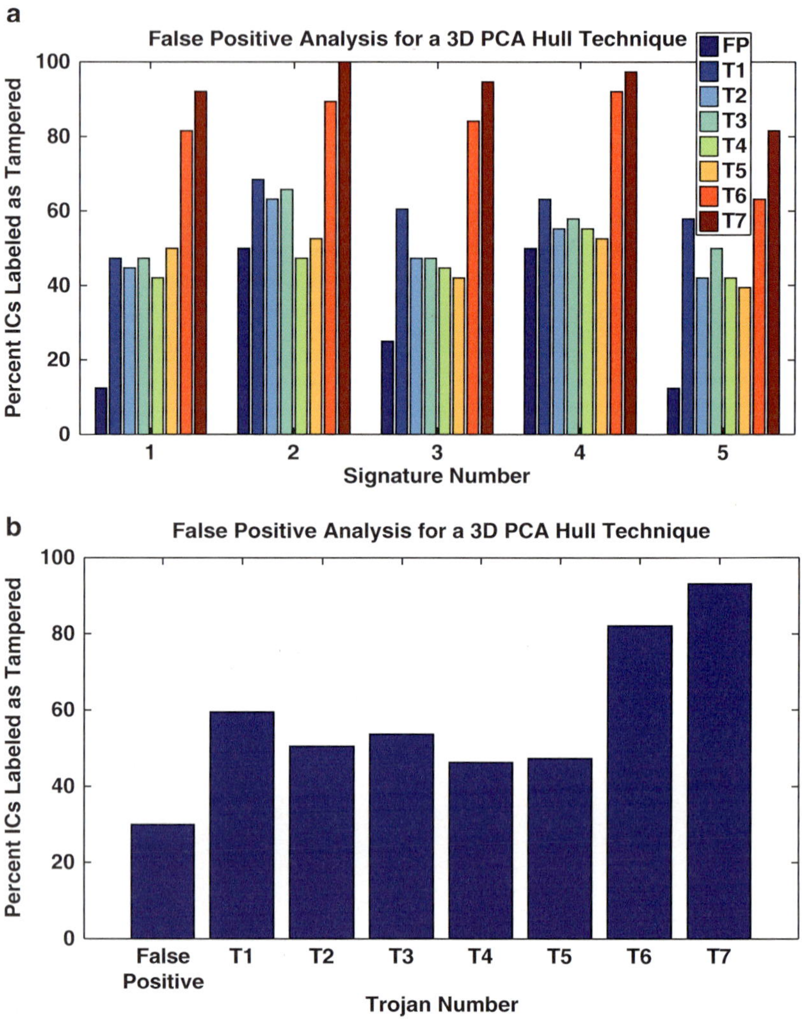

Fig. 6.28 Classification using the presented scheme and 3 dimensions. (**a**) All cases. (**b**) Mean rates

4. Determine if the CUA point falls within the hull. If it is within the boundaries of the hull it is considered Trojan-free.

To examine the performance of this classification scheme, the data is organized into five cases in which 8 of the 38 functioning ICs are randomly selected to represent Trojan-free chips to be authenticated and the remaining ICs are used to build the golden signature. All 38 ICs are used as Trojan-inserted chips under authentication.

The classification scheme was tested using both 2 and 3 dimensional hulls using the same subset cases for both hull types. The percent chips labeled as Trojan-inserted are shown for each case using both 2 and 3 dimensions in Figs. 6.27a and 6.28a respectively. "FP" indicates the number of Trojan-free chips that were incorrectly classified. For both 2 and 3 dimensions, the behavior varies among the randomly selected cases. Thus, for clarity, the average rates among all cases are shown in Figs. 6.27b and 6.28b. For both the 2 and 3 dimensional schemes, the false positive rates are lower than the detection rates for even the smallest Trojans in the experiment. For Trojans T1–T5 the detection rates are under 50%. This is unsurprising since these Trojans consisting of fewer than 130 transistors were intentionally designed to determine and emphasize the limitations of this technique.

For the larger Trojans, the detection rates are as high as 60–70% for the 2 dimensional case and 80–90% for the 3 dimensional case. Notably, the percent ICs labeled Trojan-inserted tends to be higher for the 3 dimensional case, indicating sensitivity is related to the number of dimensions used. However, the three-dimensional case also achieves a higher ratio of detection rate to false positive rate for some cases.

These results demonstrate that the ring oscillator network scheme and the presented classification scheme can adequately separate Trojan-inserted designs from the Trojan-free designs despite the presence of obfuscating process variations. Although intra-die and inter-die variations introduce roughly 8% and 17% variations in RO frequencies, compared to the 1–3% change induced by the inserted Trojans, this technique successfully classifies ICs by exploiting the spatially correlated nature of process variations.

6.7 Summary

In this chapter, an effective Trojan detection framework is presented, which combines an on-chip structure with off-chip current measurements. This technique has the capability of detecting very small Trojans with very little contribution to circuit transient current. Statistical analysis distinguishes the effects of hardware Trojans from process variations. The experimental results on 45nm FPGAs and on 90nm test chips demonstrated that this approach is very effective at identifying Trojan-inserted ICs.

References

1. Xuehui Zhang, Andrew Ferraiuolo, and Mohammad Tehranipoor, "Detection of Trojans using a Combined Ring Oscillator Network and Off-chip Transient-Power Analysis," *ACM Journal on Emerging Technologies in Computing Systems (JETC)*, 2012.
2. J. M. Rabaey, A. Chandrakasan, and B. Nikolic, "Digital Integrated Circuits: A Design Perspective (2nd Edition)," Prentice Hall, ISBN: 0-13-090996-3, 2003.

3. S. Zhao, K. Roy, and C. Koh, "Frequency Domain Analysis of Switching Noise on Power Supply Network," *Tecnical Reports*, 2000.
4. I. T. Jolliffe, "Principal Component Analysis (2ed Edition)," Springer, pp. 150–165, 2002.
5. F. P. Preparata and S. J. Hong, "Convex Hulls of Finite Sets of Points in Two and Three Dimensions," *Commun. ACM*, vol. 20, no. 2, pp. 87–93, 1977.
6. S.H.K. Embabi. "Digital BiCMOS Integrated Circuit Design." Kluwer, 1993.
7. T. Sakurai and R. Newton, "Alpha-power law MOSFET model and its applications to CMOS inverter delay and other formulas," *IEEE J. Solid-State Circuits*, vol. 25, no. 2, pp. 584–594, Apr. 1990.
8. *http://digilentinc.com/Products/Detail.cfm?NavPath=2,66,828&Prod=ADEPT2.*
9. S. Narasimhan, D. Dongdong, R. Chakraborty, S. S. Paul, F. Wolff, C. Papachristou, K. Roy, and S. Bhunia, "Multiple-parameter Side-channel Analysis: A non-invasive Hardware Trojan Detection Approach," in Proc. *IEEE HOST*, pp. 13–18, 2010.
10. *http://trust-hub.org/resources/benchmarks.*
11. D. Agrawal, S. Baktir, D. Karakoyunlu, P. Rohatgi, and B. Sunar, "Trojan Detection using IC Fingerprinting," in Proc. *IEEE Symposium on Security and Privacy (SP)*, pp. 296–310, 2007.
12. M. Banga and M. Hsiao, "A Region based Approach for the Identification of Hardware Trojans," in Proc. *IEEE HOST*, pp. 40–47, 2008.
13. F. Wolff, C. Papachristou, S. Bhunia, and R. S. Chakraborty, "Towards Trojan-free Trusted ICs: Problem Analysis and Detection Scheme" in Proc. *Design, Automation and Test in Europe (DATE)*, pp. 1362–1365, 2008.

Chapter 7
Design Vulnerability Analysis

Due to globalization, designs are vulnerable to Trojan insertion at any stage of their development. Third-party intellectual properties (IPs) are pervasive throughout high-level implementation. Their details are not delivered most of the time; therefore, IPs may stealthily deviate from the designers' determined specifications. Rouge designers or manufacturers far from the design house can tamper with synthesized designs by inserting extra gates or deleting some gates. It is also possible for circuit masks to be manipulated in foundries, changing these circuits' characteristics. Consequently, analyzing the vulnerability of a circuit to Trojan insertion at different levels is a key step towards trusted-design development [1].

Although there has been a significant amount of work on hardware Trojan detection and prevention [6–12], no systematic approach has been developed to assess a circuit's susceptibility to Trojan insertion. Adversaries target sections in a circuit with low controllability and observability and implant stealthy Trojans [13–16]. This necessitates a thorough circuit analysis to identify potential Trojan locations. Furthermore, there has been little or no work on the development of a metric to determine the difficulty of detecting a Trojan in a circuit. A comprehensive vulnerability analysis at the behavioral, gate, and layout levels is presented to quantify the difficulties of activation and observation of each circuit portion. In the following, Trojan detectability is determined based on a Trojan's activation probability and its contribution to circuit characteristics. The detectability offers a fair comparison between different Trojan detection techniques. Finally, suggestions are provided for reducing circuit susceptibility to Trojan insertion.

7.1 Vulnerability Analysis Flow

It takes many steps to produce a circuit accommodating specific requirements and functions. In a gross view, circuit functionality first determines circuit specifications including operating conditions (power, noise, frequency, etc.), environmental

M. Tehranipoor et al., *Integrated Circuit Authentication: Hardware Trojans and Counterfeit Detection*, DOI 10.1007/978-3-319-00816-5_7,

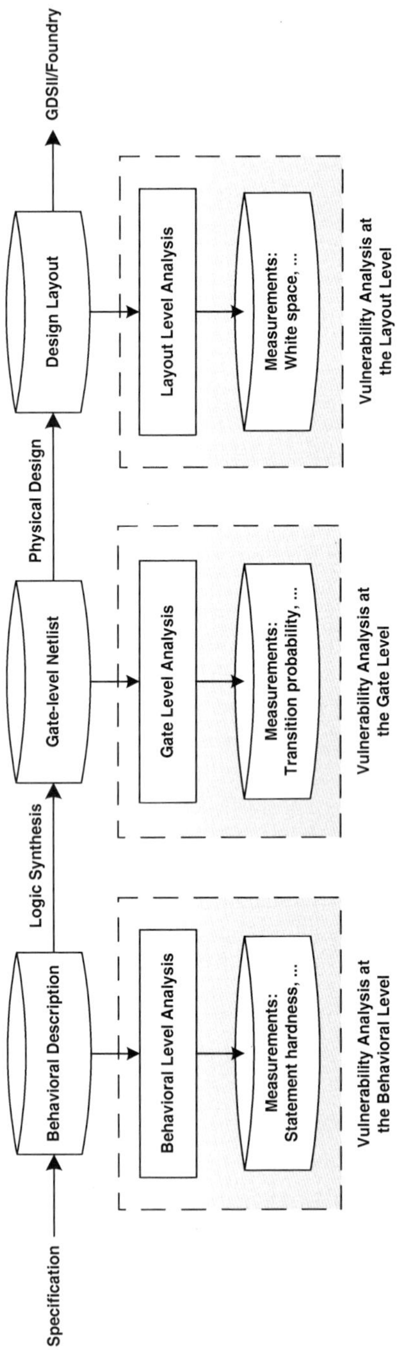

Fig. 7.1 The circuit vulnerability analysis flow

attributes (temperature, humidity, reliability, etc.), function (input–output characteristics), and other characteristics (volume, cost, price, availability, etc.) [17]. Then, a high-level behavioral implementation of the circuit is developed to realize the functional specification. Afterwards, the behavioral implementation is synthesized into a gate-level netlist. Finally, the gate-level netlist undergoes the physical design step to extract the layout of the circuit used for fabrication. Nowadays, the horizontal design integration paradigm has been widely adopted, meaning that each of these steps could be carried out by different companies distributed across the globe.

At the behavioral level, a hardware description language (HDL), such as VHDL or Verilog, is used to describe a circuit by concurrent algorithms (behavioral). Each algorithm itself is sequential, which means it consists of a set of statements that are executed one after another. An adversary can change a statement without an observable target at any output or include a new statement run under a new conditional block that gets rarely activated. At the gate level, a circuit is modeled as an interconnection of Boolean gates. A rare triggering vector of nets with low transition probabilities can be used as a triggering condition for a Trojan. Connecting Trojan payload to non-critical paths can also reduce Trojan impact on circuit performance. At the layout level, the gates are implemented as physical devices, and wire connections realize their interconnection. A chip layout determines cell placement and routing. An adversary can place Trojan cells in the layout's white spaces or cause reliability issues by widening or narrowing wires.

To effectively address Trojan prevention and detection, it is of prime importance to evaluate the susceptibility of each part of a circuit to Trojan insertion. Shown in Fig. 7.1, a comprehensive flow has been developed to perform independent design vulnerability analysis at behavioral, gate, and layout levels. Vulnerability analysis at the behavioral level begins with a circuit described in VHDL language and determines the hardness of executing each statement of code and the observability of circuit signals. At the gate level, to measure a Trojan's resiliency to power and delay side-channel analyses, the transition probability of every net and the delay of the longest path to which the net belongs are determined. At the layout level, the vulnerability analysis screens circuit layout to find possible locations for Trojan cell placement and their distributions.

7.2 Vulnerability Analysis at the Behavioral Level

At this level, a circuit is stated in the form of concurrent and sequential statements. HDL constructs, such as loop and condition blocks, direct the execution order of these statements. The circuit's data and control flows determine the hardness of executing a statement (statement hardness) and the observability of internal signals at circuit outputs. A behavioral level circuit is vulnerable to Trojan insertion when statement hardness is high or observability is low. A Trojan at the behavioral level (it is also called "a behavioral Trojan") can change a statement that is rarely executed or carry out an attack through a signal with very low observability.

7.2.1 Statement Hardness

The proposed vulnerability analysis at the behavioral level quantifies (i) statement hardness for each circuit statement and (ii) observability for each circuit signal based on a weighted value range. Adopted from the work presented in [18] to improve the accuracy of static value and branch prediction in compilers, the notation $\{W[L, U]\}$ is developed for each signal where W represents the weight of the value range, and L and U show the lower and upper limits of the value range. Circuit control and data flows determine W, L, and U for each statement. The technique creates a condition stack to track the circuit control flow. Further, it generates an individual stack for each signal to pursue the circuit data flow. While the circuit code is being parsed, observing a conditional block limiting the value range of a signal, such as loop and condition statements, pushes a new condition into the condition stack and a new weighted value range into the stack of the signal. Reciprocally, exiting a conditional block pops the condition stack and may update the stacks of signals.

A weighted value range is determined by a condition or assignment statement. The weight of a range is defined as $\left(\frac{U-L+1}{U_O-L_O+1}\right)$ where U and L are the upper and lower limits of the controlling signal at the top of the condition stack, and U_O and L_O are the declared upper and lower limits of the controlling signal. If different assignment statements to one target signal take place under different conditions of a same controlling signal in one conditional block, the weighted value ranges of the target signal would merge when exiting the condition block. Further, the statement hardness of a statement is defined as the reverse of the weight of the controlling signal of that statement (*the statement weight*).

The technique was applied to a small program, shown in Fig. 7.2, and the statement hardness of each statement and the weighted value range of controlling signals are presented in Table 7.1. Further, Fig. 7.3 shows the condition stack and stacks for controlling signals X and Y over the code execution.

The program in Fig. 7.2 consists of one sequential block between Lines 12 and 33. In Line 15 where the block begins, the condition stack is set to empty. The first statement in Line 16 executes unconditionally, and its statement hardness is set to 1. This also holds for Lines 17, 18, and 19. However, the assignment statement in Line 16 updates the weighted value range of Y to $\{1[0, 0]\}$ as the constant value 0 is unconditionally assigned to Y; i.e. $W = 1$, $L = 0$, and $U = 0$. Any new weighted value range of a signal is saved in the stack of the signal. Lines 20–29 run inside the loop defined in Line 19. Without affecting the weighted value range of Y, the loop statement in Line 19 limits the value range of the controlling signal X between 0 and 9, so $L = 0$ and $U = 9$. While $L_O = 0$ and $U_O = 15$, as declared in Line 5, the weight of the value range is $(\frac{9-0+1}{15-0+1} =)0.625$. Therefore, the hardness of statements in Lines 20, 21, and 22 are $\frac{1}{0.625}$. Represented as X_{19} in the condition stack where subscripted 19 shows the line number, the signal X is pushed to the condition stack in Line 19 and $\{0.625[0, 9]\}$ into the stack of signal X shown in Fig. 7.3. The condition statement in Line 22 again pushes the signal X (X_{22}) into the condition stack and pushes the new weighted value range $\{0.125[0, 1]\}$ into the

```
1.  LIBRARY IEEE ;
2.  USE IEEE.STD_LOGIC_1164.ALL ;
3.  ENTITY EXAMPLE IS     PORT(
4.       CLK : IN BIT ;
5.       X : IN INTEGER RANGE 0 TO 15 ;
6.       Z : OUT INTEGER );
7.  END EXAMPLE;
8.  ARCHITECTURE SIMPLE OF EXAMPLE IS
9.  SIGNAL XP : INTEGER ;
10. SIGNAL H : INTEGER ;
11. BEGIN
12.   PROCESS
13.        VARIABLE Y : INTEGER RANGE 0 TO 15 ;
14.        VARIABLE LOOPING : BOOLEAN ;
15.        BEGIN
16.        Y := 0 ;
17.        H := 0 ;
18.        LOOPING := FALSE ;
19.        FOR X IN 0 TO 9 LOOP
20.             LOOPING := TRUE ;
21.             WAIT UNTIL (CLK'EVENT) AND (CLK='1') ;
22.             IF ( X < 2 ) THEN
23.                  Y := X - 1 ;
24.             ELSE
25.                  Y := X + 1 ;
26.                  IF ( Y > 5 ) THEN
27.                       H := Y + 1 ;
28.                  END IF ;
29.             END IF ;
30.        END LOOP ;
31.        LOOPING := FALSE ;
32.        Z <= H ;
33.   END PROCESS ;
34. END SIMPLE ;
```

Fig. 7.2 A sample behavioral code

stack of signal X. The hardness of Line 23 is $\frac{1}{0.125}$, and the assignment statement in Line 23 determines a new weighted value range for Y; the weighted value range $\{0.125[-1.0]\}$ is pushed into the stack of signal Y. The ELSE statement in Line 24 in Fig. 7.2 pops from the condition stack (X_{22}) and from the corresponding stack of signal X. Afterwards, the top element of the condition stack is X_{19}, so statement hardness of Line 24 is $\frac{1}{0.625}$. Further, the ELSE statement in Line 24 again pushes X (X_{24}) to the condition stack and generates a new weighted value range $\{0.5[2, 9]\}$ and pushes it into the stack of signal X. In the following, the hardness of executing statements in Lines 25 and 26 would be $\frac{1}{0.5}$. In Line 26, a new conditional block is defined over Y inside the conditional block already defined in Line 24. Before executing Line 26, in Line 25 the assignment introduces another weighted

Table 7.1 Statement hardness and weight for the code in Fig. 7.2

Line number	Statement hardness $(1/T_W)$	Value range X	Y
15	1	–	–
16	1	–	{1[0, 0]}
17	1	–	{1[0, 0]}
18	1	–	{1[0, 0]}
19	1	–	{1[0, 0]}
20	1/0.625	{0.625[0, 9]}	{1[0, 0]}
21	1/0.625	{0.625[0, 9]}	{1[0, 0]}
22	1/0.625	{0.625[0, 9]}	{1[0, 0]}
23	1/0.125	{0.125[0, 1]}	{0.125[−1, 0]}
24	1/0.625	{0.625[0, 9]}	{0.125[−1, 0]}
25	1/0.5	{0.5[2, 9]}	{0.5[3, 10]}
26	1/0.5	{0.5[2, 9]}	{0.5[3, 10]}
27	1/0.3125	{0.5[2, 9]}	{0.3125[6, 10]}
28	1/0.5	{0.5[2, 9]}	{0.5[3, 10]}
29	1/0.625	{0.625[0, 9]}	{0.125[−1, 0], 0.5[3, 10]}
30	1	–	{0.125[−1, 0], 0.5[3, 10]}
31	1	–	{0.125[−1, 0], 0.5[3, 10]}
32	1	–	{0.125[−1, 0], 0.5[3, 10]}

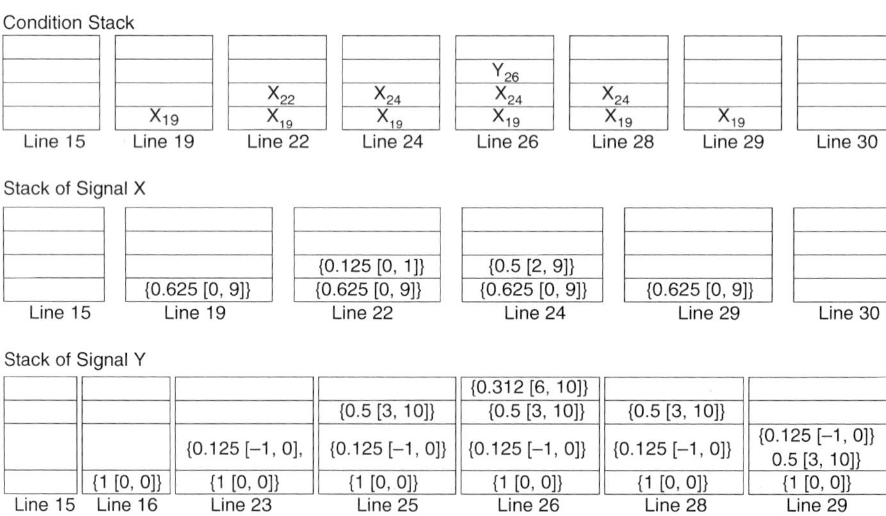

Fig. 7.3 The control flow for the code in Fig. 7.2

value range of Y, i.e. $\{0.5[3, 10]\}$, which is pushed into the stack of signal Y. The condition statement in Line 26 pushes Y (Y_{26}) into the condition stack and the weighted value range $\{0.3125[6, 10]\}$ into the stack of signal Y, so that the hardness of the statement in Line 27 would be $\frac{1}{0.3125}$. The END IF statement in Line 28 pops the stack of signal Y and Y_{26} from the condition stack, as shown in Fig. 7.3. The top element of the conditional stack would be X_{24}; therefore, the hardness of Line 28 would be $\frac{1}{0.5}$. The next END IF in Line 29 again pops the condition stack (X_{24}) and the stack of signal X. The top condition would be X_{19}, and the statement hardness of Line 29 would be $\frac{1}{0.625}$. As Y was the target of two assignments in Lines 23 and 25 under the same controlling signal of X, the two top weighted value ranges of Y are merged and saved as one element $\{0.125[-1, 0], 0.5[3, 10]\}$ at the top of the stack of signal Y. The condition stack is again popped with END LOOP in Line 30, and it becomes empty; Lines 30, 31 and 32 run unconditionally and their statement hardness would be 1.

Independent from circuit input vectors, statement hardness provides a quantitative measure for the difficulty of executing a statement. A statement with high hardness is rarely executed and its correctness cannot be fully examined by applying a limited number of test input vectors in a testbench form. In the example above, the assignment statement in Line 23 with a statement hardness of 8 is the most difficult statement in the code to be exercised, due to the small range of values in Line 22 that controls the execution of the statement. Therefore, modifications to the statement or additional statements would be hard to detect. This analysis reveals parts of a circuit that are more vulnerable to Trojan insertion. As another measure, the observability of a signal characterizes the difficulty of observing the signal through a circuit output.

7.2.2 *Observability*

A Trojan at the behavioral level can target a signal with low observability to carry out an attack. To calculate observability in parallel with statement hardness, the vulnerability analysis technique develops a weighted data graph of a circuit while it parses a circuit code. The graph shows the connectivity of the signals that appeared in the code. Figure 7.4 shows the data graph of the code in Fig. 7.2.

The nodes of the data graph are circuit signals, and directed edges show their dependency. Edges are weighted with the sum of the weights of assignment statements that contain the source node as an input and the destination node as the target signal. For example, the signal X appears in two assignment statements in Lines 23 and 25 where the target signal in each is the signal Y. The weights of the assignment statements, as shown in Table 7.1, are 0.125 and 0.5, respectively; therefore, the weight of the edge from the signal X to the signal Y is (0.125+0.5=)0.625. As another case, the weight of the edge from the signal Y to the signal H is the weight of the assignment statement in Line 27 which is 0.3125, as shown in Table 7.1. Based on the data graph, it is possible to calculate the reachability of

Table 7.2 The reachability and observability of signals for the code in Fig. 7.2

Signal name	Target signal	Reachability	Observability(T_O)
H	Z	1	1
TRUE	LOOPING	0.625	NO[1]
X	H	0.1953125	0
X	Z	0.1953125	0.1953125
X	Y	0.625	0
FALSE	LOOPING	2	NO
Y	Z	0.3125	0.3125
Y	H	0.3125	0

[1]Not Observable.

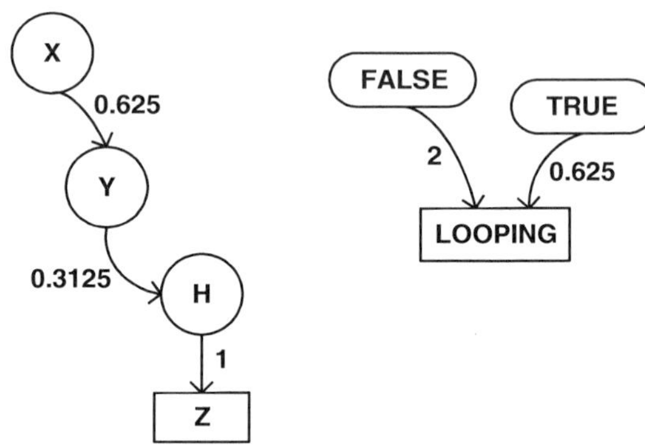

Fig. 7.4 The data graph for the code in Fig. 7.2

signals from each other and the observability of signals at circuit outputs. Table 7.2 shows the reachability and observability of signals in the circuit shown in Fig. 7.2 based on the data graph in Fig. 7.4.

The vulnerability analysis flow at the behavioral level determines the reachability of signals that can directly or indirectly reach each other. With the assignment statement in Line 32, the signal H directly reaches the signal Z with the reachability of 1, which is the weight of the edge from H to Z. As the signal Z is an output signal, the observability of H would be 1 as well. The value TRUE reaches to the signal $LOOPING$ with the reachability of 0.625. However, the signal $LOOPING$ is not observable through any circuit output signal, so the observability of TRUE is set to NO (Not Observable). The signal X reaches the signal H through the signal Y, and its reachability is $(0.625 \times 0.3125\ =)0.1953125$, but its observability is 0 as the signal H is not an output signal. However, the signal X is observable through the signal Z, which is an output signal. The reachability of the signal X to the signal Z is $(0.625 \times 0.3125 \times 1\ =)0.1953125$, and with the same value it is observable. In a similar manner, the reachability and observability of other pairs of signals can be calculated. The signal H has the highest observability, as it is directly assigned to the output signal Z. On the other hand, the signal X is observable with the minimum value of 0.1953125 through the output signal Z. As the values for TRUE and FALSE are only assigned to the signal $LOOPING$, and the signal $LOOPING$ is not an output signal, TRUE and FALSE values are not observable (NO).

The statement hardness of each statement in a code and the reachability of circuit's signals can be determined by the vulnerability analysis flow at the behavioral level. The statement hardness exposes parts of the code that are more vulnerable to Trojan insertion. Further, it is possible to obtain the weighted value range of circuit signals at any line of the code. The information can then be used to determine the root of high statement hardness. The flow also makes it possible to calculate the reachability of signals from each other and their observability through an output signal. The reachability analysis determines to what extent a signal can impact a target signal, and the observability analysis indicates how difficult it is to monitor an internal signal through an output signal.

The vulnerability analysis flow consists of two main steps. In the first step, as a code is being parsed, the hardness of each statement is calculated and a circuit graph is developed. In the following step, the reachability and observability of signals are calculated using the circuit graph. The complexity of the first step is $O(L)$ where L is the number of statements. Suppose S is the number of signals, the complexity of the second step would be $O(S \times S)$. Therefore, the complexity of the vulnerability analysis algorithm is $O(L) + O(S^2)$. It is expected that the number of signals in a circuit is much less than the number of statements ($S \ll L$); thus the complexity of the vulnerability analysis algorithm is approximated to $O(L)$.

7.2.3 Trojans Insertion at the Behavioral Level

Behavioral Trojans may target signals with low observability and lay where statement hardness is high to reduce their detection probability. To comparatively determine the difficulty of detecting behavioral Trojans across different circuits, here, behavioral Trojan detectability is defined based on statement weight and observability measures. For each Trojan statement, statement weight (T_W) and the observability of its target signal (T_O) are determined, and the Trojan statement detectability is then defined as $T_{Detectability} = T_W \times T_O$.

As an example, three Trojans were inserted in the code presented in Fig. 7.2 by changing its assignment statements, as shown in Fig. 7.5. Table 7.3 shows their detectability. The first behavioral Trojan (TjB-Loc1, where B represents "behavioral level") was realized by manipulating the statement in Line 20 and assigning FALSE rather than TRUE to the signal $LOOPING$. The weight of statement (T_W) is 0.625, according to Table 7.1, and the observability of the target signal (T_O), the signal $LOOPING$, is 0 based on Table 7.2. Therefore, the detectability of Trojan TjB-Loc1 would be $TjB - Loc1_{Detectability} = T_W \times T_O = 0.625 \times 0 = 0$. By modifying the assignment statement in Line 23, the second behavioral Trojan (TjB-Loc2) was implemented. The statement weight and the observability of the

```
LIBRARY IEEE ;
USE IEEE.STD_LOGIC_1164.ALL ;
ENTITY EXAMPLE IS     PORT(
      CLK : IN BIT ;
      X : IN INTEGER RANGE 0 TO 15 ;
      Z : OUT INTEGER );
END EXAMPLE;
ARCHITECTURE SIMPLE OF EXAMPLE IS
SIGNAL XP : INTEGER ;
SIGNAL H : INTEGER ;
BEGIN
  PROCESS
       VARIABLE Y : INTEGER RANGE 0 TO 15 ;
       VARIABLE LOOPING : BOOLEAN ;
       BEGIN
       Y := 0 ;
       H := 0 ;
       LOOPING := FALSE ;
       FOR X IN 0 TO 9 LOOP
            LOOPING := FALSE ;   -- Trojan 1 (TjB-Loc1)
            WAIT UNTIL (CLK'EVENT) AND (CLK='1') ;
            IF ( X < 2 ) THEN
                Y := X + 1 ;            -- Trojan 2 (TjB-Loc2)
            ELSE
                Y := X + 1 ;
                IF ( Y > 5 ) THEN
                     H := Y + 3 ;   -- Trojan 3 (TjB-Loc3)
                END IF ;
            END IF ;
       END LOOP ;
       LOOPING := FALSE ;
       Z <= H ;
  END PROCESS ;
END SIMPLE ;
```

Fig. 7.5 A behavioral-Trojan inserted code for the code in Fig. 7.2

Table 7.3 Trojan detectability at the behavioral level for the code in Fig. 7.5

Trojan	T_W	T_O	$T_{Detectability}$
TjB-Loc1	0.625	0	0
TjB-Loc2	0.125	0.3125	0.0390625
TjB-Loc3	0.3125	1	0.3125

target signal were 0.125 and 0.3125, respectively; therefore, $TjB-Loc2_{Detectability} = T_W \times T_O = 0.125 \times 0.3125 = 0.0390625$. The last behavioral Trojan (TjB-Loc3) was carried out by changing the assignment statement in Line 27. The detectability of TjB-Loc3 is $TjB - Loc3_{Detectability} = T_W \times T_O = 0.3125 \times 1 = 0.3125$.

The lower a Trojan detectability ($T_{Detectability}$) is, the harder it is to detect a Trojan. In the above example, TjB-Loc1 was the hardest Trojan to detect with the lowest detectability of 0 because the target signal, the signal $LOOPING$, was not observable through any output signal. On the other hand, the target signals for TjB-Loc2 and TjB-Loc3 reached the output signal Z. As the statement weight and observability of TjB-Loc3 were larger than those for TjB-Loc2, TjB-Loc3 was the easiest Trojan to detect. Hence, TjB-Loc3 had the highest detectability, and TjB-Loc2 was ranked second. Note that other potentially inserted Trojans (inserting new statements, loops ...) in a behavioral HDL code can be evaluated in a similar manner.

7.3 Vulnerability Analysis at the Gate Level

At the gate level, circuits are susceptible to hardware Trojans realized by the addition or deletion of gates. Gate-level Trojans can cause functional modification or parametric deviation, under rare conditions. To withdraw Trojan effects from established testing techniques [17], an adversary can exploit hard-to-detect areas (e.g. nets) in a circuit to implement a Trojan. Hard-to-detect areas are defined as areas not testable by established fault-testing techniques (stuck-at, transition delay, path delay, and bridging faults) or not having a noticeable impact on circuit side-channel signals (transient power and delay)[1].

Here, a circuit vulnerability analysis is proposed at the gate-level to identify hard-to-detect nets in a circuit, providing opportunities to insert Trojans that are very difficult to detect.

7.3.1 The Proposed Flow

Shown in Fig. 7.6, the gate-level vulnerability analysis flow performs power (transition probability), delay, and structural analyses on a circuit to extract hard-to-detect nets. Any transition inside a Trojan circuit increases the overall circuit transient power consumption; therefore, it is expected that Trojan inputs are supplied by nets with low transition probabilities to reduce activity inside the Trojan circuit, indicated as A_{Trojan}, compared with the activity of Trojan-free circuit, indicated as A_{TjFree}. The Power (Transition Probability) Analysis step in Fig. 7.6 determines the transition probability of every net in the circuit, assuming the probability of 0.5 at primary inputs and memory cells outputs as "0" or "1". Then, nets with transition probabilities below a certain threshold are considered possible Trojan inputs.

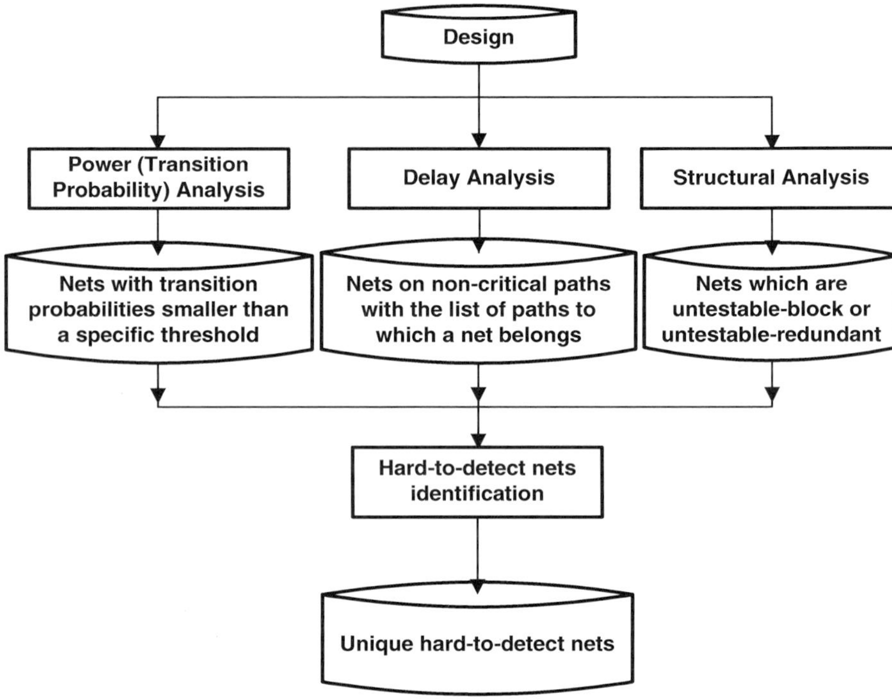

Fig. 7.6 The circuit vulnerability analysis at the gate-level

Extra capacitances induced by Trojan cells and their wiring, indicated as Trojan induced capacitance (TIC), change the timing characteristics of the nets to which a Trojan circuit taps. It is unlikely that Trojan inputs and outputs are connected to critical paths in order to keep Trojans' impact hidden. Otherwise, variations in circuit performance flag deviations from circuit specifications. The Delay Analysis step performs path delay measurement and lists the nets on non-critical paths. Here, short paths whose delays are less than 70 % of the critical paths are assumed not be measurable on silicon if the TIC on a path is less than the slack of the path. They provide great opportunities to insert Trojans since induced delay will not make the path long enough to be tested by a tester. To further reduce Trojans' impact on circuit delay characteristics, the Delay Analysis step also reports paths to which a net belongs to avoid selecting nets composing different sections of one path.

A circuit may contain redundant circuits that would not determine circuit functionality and are not testable. Since transition delay fault patterns are primarily used to test transition delay faults targeting all paths, as opposed to path delay tests that target critical paths only, here the Structural Analysis step executes the structural transition delay fault testing to find untestable-blocked and untestable-redundant nets. Untestable-redundant nets are not testable because they are masked by a redundant logic, and they are not observable through primary outputs or scan cells. Untestable-blocked nets are not controllable or observable by untestable-redundant nets. Tapping Trojan inputs to untestable nets hides Trojans' impact on delay variations.

At its end, the circuit vulnerability analysis at the gate level reports unique hard-to-detect nets; the list of untestable nets with low transition probabilities and nets with low transition probabilities on non-critical paths (i.e. short paths) while not sharing any common path. Note that when a Trojan impacts more than one path, it provides more opportunities for detection. Avoiding the use of common paths makes a Trojan's contribution to the affected path delay minimal, a presence which can be masked by process variations. This similarity blurs the line between delays induced by hardware Trojans and delays attributed to process variations. The reported nets are ensured to be untestable by structural test patterns used in production test. They also have low transition probabilities so that Trojans would negligibly affect the circuit power consumption. As the nets are chosen from non-critical paths without any shared segments, it would be extremely difficult to detect the Trojans using delay-based techniques.

The flow is applied to the Ethernet MAC 10GE circuit [19] which implements 10Gbps Ethernet Media Access Control functions. Synthesized at 90nm technology node, the Ethernet MAC 10GE circuit consists of 102,047 components, including 21,830 flip-flops.

1. The Power (Transition Probability) Analysis shows that this circuit has 102,669 nets, and 23,783 of them have a transition probability smaller than 0.1, 7003 of them smaller than 0.01, 367 of them smaller than 0.001, and 99 of them smaller than 0.0001.
2. The Delay Analysis indicates that the largest capacitance along a path (representing path delay) in the circuit is 0.065717825 PF, and there are 14,927 paths in the circuit whose path capacitances are smaller than 70% of the largest capacitance, assuming that paths longer than 70% could be tested using testers.
3. The Structural Analysis finds no untestable fault in this circuit.

By excluding nets sharing different segments of one path, there are 494 nets in the Ethernet MAC 10GE circuit considered to be areas where Trojan inputs could be used while ensuring the high difficulty of detection based on side-channel and functional test techniques.

7.3.2 Trojan Insertion at the Gate Level

Trojans' impact on circuit characteristics depends on their implementation. Trojan inputs tapped from nets with higher transition probabilities would aggrandize switching activity inside a Trojan circuit and increase the contribution of the Trojan into circuit power consumption. Furthermore, Trojans might affect circuit delay characteristics due to additional capacitance induced by extra routing and Trojan gates. To quantitatively determine the difficulty of detecting a gate-level Trojan, a procedure is developed to determine the detectability of gate-level Trojans based on their impact on delay and power across different circuits. Since it is based on induced variations by a Trojan in side-channel signals, the detectability can establish a fair comparison among different hardware Trojan detection techniques.

Trojan detectability is determined by (1) the number of transitions in the Trojan circuit and (2) extra capacitance induced by Trojan gates and routing. This metric is designed to be forward-compatible with new approaches to Trojan detection by introducing a new variable, for example, a quantity related to the electromagnetic field. The quantities are each normalized by a related circuit-dependent quantity.

Transitions in a Trojan circuit reflect Trojan contribution to circuit power consumption. The number of transitions in the Trojan circuit (A_{Trojan}) divided by the Trojan circuit size (S_{Trojan}) is normalized by the number of transitions in the Trojan-free circuit (A_{TjFree}) divided by the Trojan-free circuit size (S_{TjFree}). The number of cells in a circuit is considered the circuit size. Trojan impact on circuit delay characteristic is represented by measuring the added capacitance by the Trojan. The Trojan-affected path with the largest capacitance in the corresponding Trojan-free circuit is considered as C_{TjFree}, and the added capacitance as Trojan-induced capacitance (TIC), and then TIC is normalized by C_{TjFree}.

The detectability of a Trojan ($T_{Detectability}$) at the gate-level is calculated as follows:

1. Apply random inputs to the Trojan-free circuit and obtain the number of transitions in the circuit (A_{TjFree}).
2. Apply the same random vectors to the Trojan-inserted circuit and obtain the number of transitions in the Trojan circuit (A_{Trojan}).
3. Perform the Delay analysis on Trojan-free and Trojan-inserted circuits.
4. Obtain the list of paths whose capacitances are changed by the Trojan.
5. Determine the Trojan-affected path with the largest capacitance in the corresponding Trojan-free (C_{TjFree}) and the added capacitance (TIC).
6. Form the vector $(A/B, C/D)$ where $A = A_{Trojan}/S_{Trojan}$, $B = A_{TjFree}/S_{TjFree}$, $C = TIC$, and $D = C_{TjFree}$, and compute the vector magnitude as the Trojan detectability. Note that Trojan detectability represents the level of detection difficulty for a Trojan.

As an example, a comparator Trojan, shown in Fig. 7.7, was inserted at four different locations (TjG-Loc1, TjG-Loc2, TjG-Loc3, and TjG-Loc4 (G represents "gate level")) in the Ethernet MAC 10GE circuit, and Table 7.4 shows their detectability.

The Ethernet MAC 10GE circuit consists of 102,047 cells while the Trojan size with 12 cells is only about 0.011% of the entire circuit. TjG-Loc4 experienced the largest switching activity and induced relatively high TIC. It was expected that TjG-Loc4 would be the easiest Trojan to detect due to more impact on circuit side-channel signals, and the results confirm that the detectability of TjG-Loc4 was higher than the others. Although the induced capacitance by TjG-Loc2 was more than the capacitance induced by TjG-Loc1, TjG-Loc1 made a more significant contribution to circuit switching activity. Therefore, TjG-Loc1 had the second

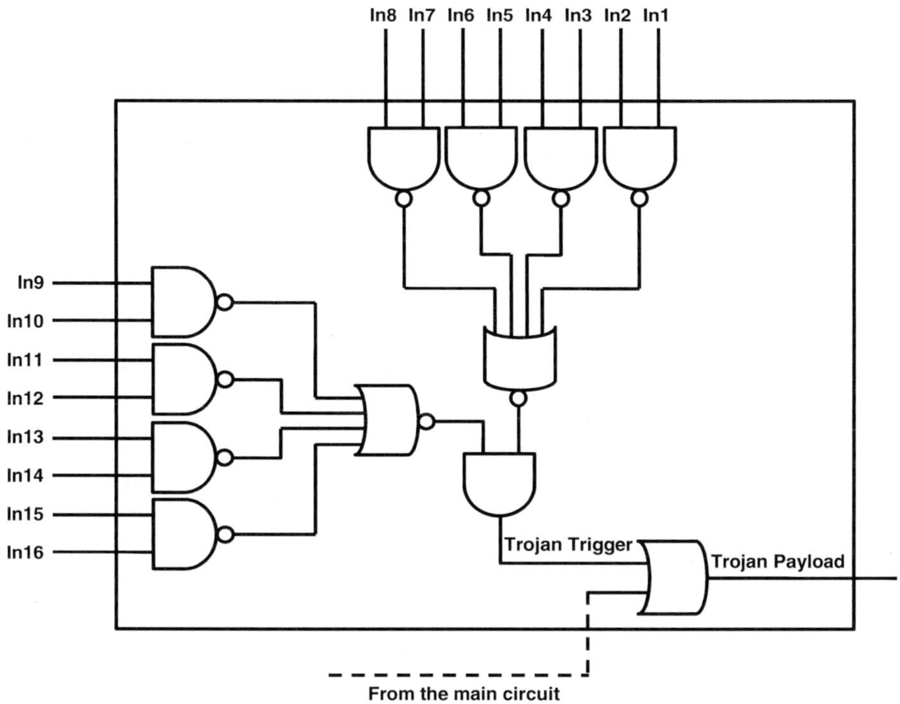

Fig. 7.7 A comparator Trojan

Table 7.4 The detectability of the comparator Trojan placed at four different locations in Ethernet MAC 10GE circuit

Trojan	A_{TjFree}	S_{TjFree}	A_{Trojan}	S_{Trojan}	$TIC(pF)$	$C_{TjFree}(pF)$	$T_{Detectability}$
TjG-Loc1	106,664,486	102,047	10,682	12	0.000286935	0.041358674	0.851659
TjG-Loc2	106,664,486	102,047	4,229	12	0.004969767	0.072111502	0.344132
TjG-Loc3	106,664,486	102,047	3,598	12	0.005005983	0.049687761	0.304031
TjG-Loc4	106,664,486	102,047	13,484	12	0.004932996	0.052602269	1.079105

largest detectability after TjG-Loc4. Among TjG-Loc2 and TjG-Loc3, though TjG-Loc3 had slightly larger induced capacitance, TjG-Loc2 experienced more switching activity. The two Trojans had close detectability where TjG-Loc2 ranked above and TjG-Loc3 remained the hardest Trojan to detect with the lowest Trojan detectability.

As another example, a different comparator Trojan with 16 inputs was inserted at two different locations, TjG-Loc5 and TjG-Loc6 in s38417 circuit [20], and Table 7.5 shows their detectability. Synthesized at 90nm technology node, s38417 consists of 5,329 components.

Table 7.5 The detectability of the comparator Trojan placed at two different locations in s38417

Trojan	A_{TjFree}	S_{TjFree}	A_{Trojan}	S_{Trojan}	$TIC(pF)$	$C_{TjFree}(pF)$	$T_{Detectability}$
TjG-Loc5	2,717,682	5,329	59	11	0.004167929	0.032341219	0.1393
TjG-Loc6	2,717,682	5,329	1,328	11	0.005313744	0.030518107	0.4108

1. The Power (Transition Probability) Analysis shows that the circuit has 5,668 nets, of which 36 nets have a transition probability of less than 0.00000001, 68 nets with less than 0.0001, and 219 nets with less than 0.001.
2. The Delay Analysis indicates that the largest capacitance along a path (representing path delay) in the circuit is 0.050146392 PF,
3. The Structural Analysis finds no untestable fault in this circuit.

While TjG-Loc5 induced larger Trojan induced capacitance, TjG-Loc6 impacted a path with larger capacitance, and the switching activity of TjG-Loc6 circuit is considerably larger. Therefore, $T_{Detectability}$ of TjG-Loc6 is larger than the $T_{Detectability}$ of TjG-Loc5, and TjG-Loc5 would be harder to detect.

As $T_{Detectability}$ is based on normalized measures, it makes it possible to compare the detectability of different Trojans across different circuits. In the above examples, Trojans TjG-Loc1 to TjG-Loc4 are inserted in Ethernet MAC 10GE circuit and Trojans TjG-Loc5 and TjG-Loc6 in s38417 circuit. Their detectability indicated that although s38417 circuit is considerably smaller than Ethernet MAC 10GE circuit, the detectability of TjG-Loc5 and TjG-Loc6 are lower than all Trojans inserted in Ethernet MAC 10GE.

7.4 Vulnerability Analysis at the Layout Level

A circuit layout carries information about cell placement and routing. A layout tool locates and connects all circuit components so that circuit requirements, such as performance, size, and manufacturability, are all met. This would usually leave unused spaces (white spaces) in a circuit layout that can potentially be used by an adversary to insert Trojan cells.

The distribution of Trojan cells across a circuit layout is a deterministic factor in Trojan impact on circuit side-channel signals. Trojan cells placed tightly in a particular area could have more impact on circuit power consumption as there would be greater localized switching activity in the area. On the other hand, the loose distribution of Trojan cells requires long wire connections for Trojan inputs and outputs and between Trojan cells. Hence, loose Trojans could affect the circuit performance or delay distribution.

To analyze the vulnerability of a circuit layout to Trojan insertion, a flow is developed to screen the circuit layout and determine possible locations for Trojan cells. The flow accepts a Design Exchange Format (DEF) file which represents the physical layout of a circuit in an ASCII format. The distribution of circuit cells and white spaces across the circuit layout are obtained, and potential locations for circuit cells placement are then determined.

To evaluate the flow, first, physical design was performed on b19 benchmark [21] using Synopsys IC Compiler at 90nm technology node. Then, its DEF file was obtained and the vulnerability analysis at the layout level was performed. The b19 circuit consists of 62,835 cells, and Fig. 7.8a,b respectively show the distribution of circuit cells and the distribution of white spaces as a unit of INVX0 (the smallest gate in SAED_EDK90nm library [22]) across the layout. The layout is divided into tiles with the size of the largest cell in the design library.

The results indicated that the densest area of the layout has 106.67 units of INVX0 cells around the center of the layout, and the average density is 57.01 units of INVX0. With an average size of 31.90 units of INVX0, white spaces are more prevalent in areas close to the layout boundaries, and there are plenty of areas with low density across the circuit layout.

White spaces adjacent to areas with high density are suitable places to insert Trojans resilient to power-based Trojan detection techniques. The high power consumption of dense areas can mask the small contribution of Trojan cells to circuit power consumption. To address excessive power consumption, white spaces may be filled with decoupling capacitances by chip designers. Even if there are not enough white spaces available, an adversary can open space for Trojan cells by carefully examining the layout and removing some decoupling capacitances. The availability of white spaces across a circuit layout also makes it easier to insert Trojans resilient to delay-based Trojan detection techniques. Placing Trojan cells close to their driving cells reduces induced capacitances due to Trojan wire connections.

The detectability of functional Trojans at the layout level, similar to Trojans at the gate-level, can be determined by considering Trojan-induced delay and the number of transitions in Trojan and main circuits. However, at the layout level it is possible to incorporate Trojan wiring capacitance and switching density around Trojan cells in detectability calculations.

In addition to functional Trojans, Trojans at the layout level can be realized by disregarding layout design rules (parametric Trojans). Wires running a long distance in parallel might be widened to increase circuit susceptibility to crosstalk. Narrowing wires could also reduce circuit reliability as the wires might melt due to the high current flow during high switching activity periods. While layout-level Trojans could severely affect circuit functionality or significantly reduce circuit reliability, off-line and off-chip testing might not effectively target these Trojans.

7.5 Alleviating Circuit Vulnerabilities

The vulnerability analysis showed that a circuit at any stage of development is susceptible to Trojan insertion. At the behavioral level, a Trojan can be inserted into parts with a high statement hardness and can carry out an attack through signals with

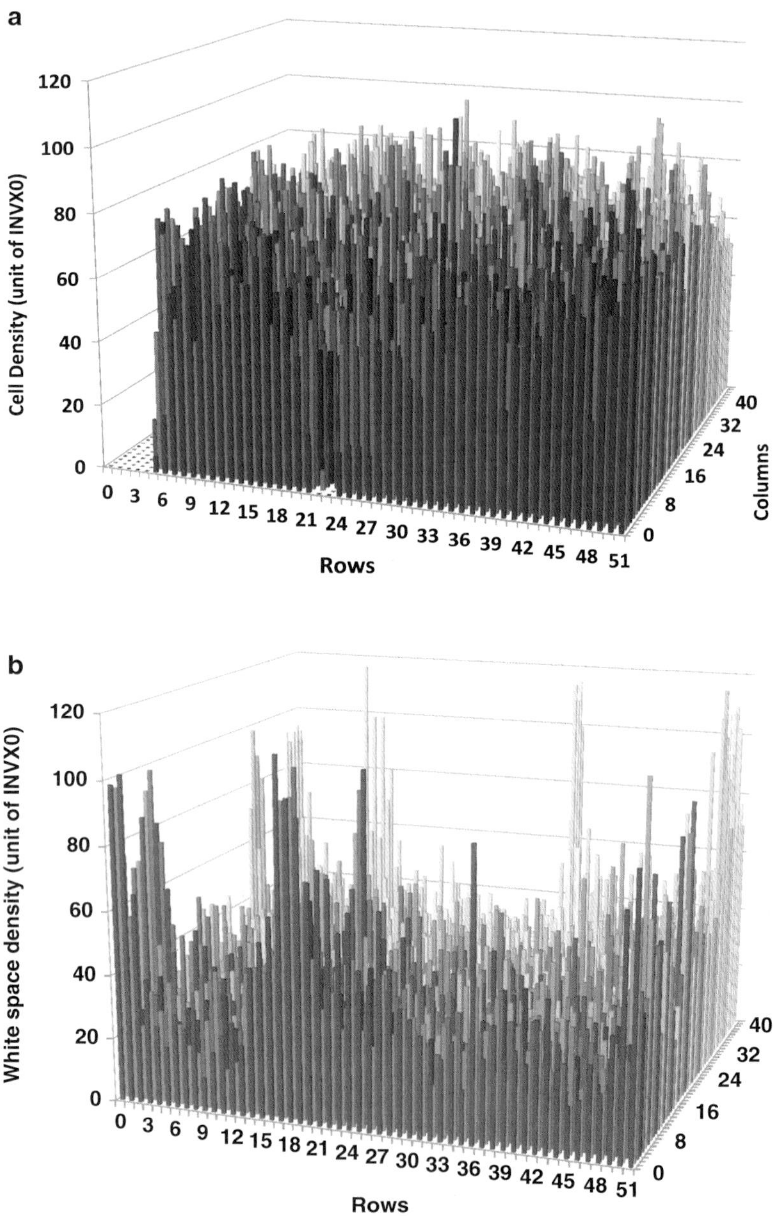

Fig. 7.8 Layout vulnerability analysis for b19. (**a**) The distribution of cells across b19 layout. (**b**) White space density across b19 layout

low observability. To reduce Trojans' detection probabilities at the gate level, Trojan triggers are supplied by nets with low transition probabilities without impacting critical paths. White spaces in a circuit layout can house Trojan cells without any impact on the size of circuit layout.

Each level of a circuit requires particular measures to prevent Trojan insertion and to increase Trojan detection probabilities. To increase the probability of detecting behavioral Trojans, circuit statements with high statement hardness and signals with low observability should first be distinguished. Then, some techniques should be employed to ease executing statements with high statement hardness and to facilitate monitoring signals with low observability. For example, adding extra control statements accessible though primary inputs might make the execution of statements with high statement hardness more frequent. Passing low observable signals to primary outputs under a new set of statements and conditions could limit an adversary's ability to insert Trojans.

At the gate level, Trojan triggers with low activation probabilities limit Trojan contribution to circuit side-channel signals and make the logic testing less effective. The detection probability of gate-level Trojans can be increased by removing rare-triggering conditions by increasing the transition probabilities of all circuit nets above a certain threshold. It is possible to increase nets' transition probabilities by providing immediate access to the internal parts of a circuit through primary inputs, scan flip-flops, or dummy cells [8]. Without affecting circuit functionality and performance, monitoring circuits can also be embedded into a circuit to improve testing [2–5].

Layout-level Trojans can take advantage of white spaces in a circuit layout to place Trojan cells without needing to change the size of circuit area. Even it is possible for an adversary to implant a Trojan causing circuit reliability issues. To prevent Trojan cell placement, white spaces in a circuit layout should be testable. For example, they can be filled with a specific logic easily testable to ensure white spaces are not used for Trojan implementation. Another approach would be to minimize or eliminate white spaces in the circuit, filling white spaces with dummy logic.

7.6 Summary

There is currently no systematic approach to evaluate the susceptibility of a circuit to Trojan insertion. This chapter presented a vulnerability analysis at the behavioral, gate, and layout levels to quantitatively determine the difficulty of detecting Trojans. At the behavioral level, the hardness of executing any statement and the observability of any signal were analyzed. Then, the detectability of behavioral Trojans was determined based on their statement weight and the observability of their target signals. The vulnerability analysis at the gate level calculated the transition probabilities of circuit nets and the capacitance of paths to which the nets belong. In the following, gate-level Trojans were ranked based on their activation probability and induced delay. White spaces of a circuit layout could be used to place Trojan cells without any impact on circuit size and circuit form factor. The vulnerability analysis at the layout level screened a layout to determine the distribution of white spaces and circuit cells. The distribution information could determine possible areas where Trojan cells were placed.

References

1. M. Tehranipoor and F. Koushanfar, "A Survey of Hardware Trojan Taxonomy and Detection," *IEEE Design and Test of Computers*, pp. 10–25, 2010.
2. M. Li, A. Davoodi, and M. Tehranipoor, "A sensor-assisted self-authentication framework for hardware Trojan detection," in Proc. of the *IEEE Design, Automation & Test in Europe Conference & Exhibition (DATE12)*, pp. 1331–1336, 2012.
3. C. Lamech, J. Aarestad, J. Plusquellic, R. Rad, and K. Agarwal "REBEL and TDC: Two embedded test structures for on-chip measurements of within-die path delay variations," in Proc. of the *IEEE/ACM International Conference on Computer-Aided Design (ICCAD11)*, pp. 170–177, 2011.
4. X. Zhang and M. Tehranipoor, "RON: an on-chip ring oscillator network for hardware Trojan detection," in Proc. of the *IEEE Design, Automation & Test in Europe Conference & Exhibition (DATE11)*, pp. 1–6, 2011.
5. L. Kim and J. Villasenor "A system-on-chip bus architecture for thwarting integrated circuit Trojan horses," *IEEE Transaction on Very Large Scale Integration (TVLSI) SYSTEMS*, Volume. 19, Issue. 10, pp. 1921–1926, 2011.
6. M. Banga and M.S. Hsiao, "ODETTE: A issuen-scan design-for-test methodology for Trojan detection in ICs," in Proc. of the *IEEE International Symposium on Hardware-Oriented Security and Trust (HOST11)*, pp. 18–23, 2011.
7. H. Salmani and M. Tehranipoor, "Layout-aware switching activity localization to enhance hardware Trojan detection," *IEEE Transactions on Information Forensics and Security (TIFS)*, Volume: 7, Issue: 1, pp. 76–87, 2012.
8. H. Salmani M. Tehranipoor, and J. Plusquellic, "A novel technique for improving hardware Trojan detection and reducing Trojan activation time," *IEEE Transactions on Very Large Scale Integration (TVLSI) Systems*, Volume: 20, Issue: 1, pp. 112–125, 2012.
9. J. Aarestad, D. Acharyya, R. Rad, and J. Plusquellic, "Detecting Trojans through leakage current analysis using multiple supply pad I_{DDQ}s," *IEEE Transactions on Information Forensics and Security (TIFS)*, Volume: 5 , Issue: 4, pp. 893–904, 2010.
10. J. Zhang, H. Yu, and Q. Xu, "HTOutlier: Hardware Trojan detection with side-channel signature outlier identification," in Proc. of the *IEEE International Symposium on Hardware-Oriented Security and Trust (HOST12)*, pp. 55–58, 2012.
11. S. Wei, S. Meguerdichian, and M. Potkonjak, "Malicious circuitry detection using thermal conditioning," *IEEE Transactions on Information Forensics and Security (TIFS)*, Volume: 6, Issue: 3, pp. 1136–1145, 2011.
12. C. Lamech and J. Plusquellic, "Trojan detection based on delay variations measured using a high-precision, low-overhead embedded test structure," in Proc. of the *IEEE International Symposium on Hardware-Oriented Security and Trust (HOST12)*, pp. 75–82, 2012.
13. Y. Shiyanovskii, F. Wolff, A. Rajendran, C. Papachristou, D. Weyer, and W. Clay, "Process reliability based Trojans through NBTI and HCI effects," in Proc. of the *NASA/ESA Conference on Adaptive Hardware and Systems (AHS10)*, pp. 215–222, 2010.
14. X. Wang, S. Narasimhan, A. Krishna, T. Mal-Sarkar, and S. Bhunia, "Sequential hardware Trojan: Side-channel aware design and placement," in Proc. of the *IEEE 29th International Conference on Computer Design (ICCD11)*, pp. 297–300, 2011.

15. G. T. Becker, A. Lakshminarasimhan, L. Lang, S. Srivathsa, V.B. Suresh, and W. Burelson, "Implementing hardware Trojans: experiences from a hardware Trojan challenge," in Proc. of the *IEEE 29th International Conference on Computer Design (ICCD11)*, pp. 301–304, 2011.
16. X. Zhang, N. Tuzzio, and M. Tehranipoor, "Red Team: design of intelligent hardware Trojans with known defense schemes," in Proc. of the *IEEE 29th International Conference on Computer Design (ICCD11)*, pp. 309–312, 2011.
17. M. Bushnell, V. Agrawal, *Essentials of electronic testing for digital, memory, and mixed-signal VLSI circuits,* Kluwer Academic Publishers, 2002.
18. W. H. Harrison, "Compiler analysis of the value ranges for variables," *IEEE Transactions on Software Engineering,* Volume: SE-3, Issue: 3, pp. 243–250, 1977.
19. Ethernet 10GE MAC, *http://opencores.org/project,xge_mac.*
20. ISCAS89 Benchmark, *http://www.pld.ttu.ee/~maksim/benchmarks/iscas89/verilog/.*
21. ITC99 Benchmark, *http://www.cerc.utexas.edu/itc99-benchmarks/bench.html.*
22. Synopsys 90nm generic library for teaching IC design, *http://www.synopsys.com/Community/UniversityProgram/Pages.*

Chapter 8
Trojan Prevention: Built-In Self-Authentication

The complexity and cost of IC fabrication significantly increases as the technology for integrated circuits (ICs) shrinks to very deep sub-micron levels. As companies increasingly outsource design and fabrication for economic reasons, it has become easier for adversaries to implant a Trojan by modifying the layout of a circuit during GDSII development and fabrication. A circuit's unused spaces provide great opportunities for Trojan insertion.

Although there has been considerable effort made towards improving Trojan detection [2–10], less attention has been paid to Trojan prevention. To prevent Trojan insertion during physical design, built-in self-authentication (BISA) technique is proposed, which can fill *all* unused spaces in a circuit layout with functional standard cells instead of non-functional filler cells. BISA can test the functionality of all functional filler cells automatically with low overhead; therefore, BISA is able to prevent hardware Trojan insertion. Additionally, BISA's impact on the original circuit is negligible because there is no connection between the BISA circuit and the original circuit, and they do not work simultaneously. In the meantime, the added BISA cells can provide decoupling capacitances to minimize voltage drop when the original circuit is working. Moreover, BISA is immune to different types of attacks. Changing or removing any gate belonging to BISA can be detected by BISA itself. BISA insertion can be automated so that designers do not need any knowledge of Trojans. Finally, BISA eliminates the opportunity for an untrusted GDSII developer and foundry to add any malicious circuitry; however, they may still be able to carry out their intentions by removing circuit gates and adding their own cells. But this is an easier problem to address through detection techniques because of the changes to the circuit's functionality.

M. Tehranipoor et al., *Integrated Circuit Authentication: Hardware Trojans and Counterfeit Detection*, DOI 10.1007/978-3-319-00816-5_8,

8.1 BISA Structure and Insertion Flow

Placement tools are typically conservative to limit the density and spread cells evenly, in order to assure routability. This often leaves small white spaces between cells; it is impossible to fill 100 % of the area with regular standard cells in VLSI designs. After completing placement and routing, designers usually fill these unused spaces with filler cells or decoupling capacitor (DECAP) cells to reduce the design rule check (DRC) violations created by the base layers and ensure power rail connection [11]. However, filler cells do not have functionality. If designers want to make changes, known as Engineering Change Orders (ECO), filler cells can be deleted and the empty spaces can be utilized for new cells. On the other hand, intelligent attackers have seen this opportunity to identify and remove some filler cells for Trojan insertion because removing these non-functional filler cells does not change the original functionality. If attackers redesign the original layout for Trojan insertion instead of adding Trojan gates in unused spaces, moving gates' locations or altering wire interconnections will result in significant changes to side-channel signals. These can be detected much more easily by delay-based and power-based techniques [1].

The principal idea of BISA is to fill all unused spaces with functional standard cells, named BISA cells, instead of conventional non-functional filler cells. These BISA cells are connected together to form a combinational circuit that is independent from the original circuits. By testing functions of the combinational circuit composed of BISA cells, designers would be able to find out if these cells are modified or not after fabrication. The BISA circuit is designed so that stuck-at patterns can test all its gates, thus any change in BISA cells can be detected. Furthermore, BISA cells are the same as the standard cells the circuit uses, so identifying these cells will be extremely difficult, if not impossible. In this section, BISA structure and how BISA is applied to conventional bottom-up hierarchical ASIC designs, including single module design and System-on-Chip (SOC) design, are explained.

8.1.1 BISA Structure and Function

To test the inserted filler cells with low overhead, a similar methodology to the Logic Built-In Self-Test (LBIST) system in VLSI testing is adapted. Basically, the BISA, as in LBIST, is composed of three parts: the BISA circuit under test, the Test Pattern Generator (TPG), and the Output Response Analyzer (ORA), as shown in Fig. 8.1. The BISA circuit under test is composed of all BISA cells which are inserted into unused spaces during physical design. Generally, it is easier to obtain higher test coverage with a smaller combinational circuit with fewer gates. In order to increase its stuck-at fault test coverage, the BISA circuit is divided into a number of small combinational logic blocks, called BISA blocks. Each BISA

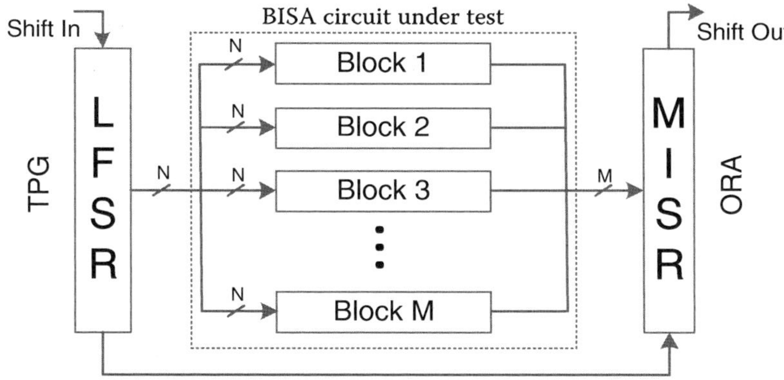

Fig. 8.1 The BISA structure

Table 8.1 Operation modes in a design with BISA

		Authentication mode	
	Normal mode	Shift mode	Test mode
Original circuit	Working	Idle	Idle
BISA circuit	Idle	Shift seed/signature	Testing BISA

block can be considered as an independent combinational logic block. The TPG generates test vectors that are shared by all BISA blocks. The ORA will process the outputs of all BISA blocks and generate a signature. In Fig. 8.1, a linear-feedback-shift-register (LFSR) is used as the TPG and a multiple-input-signature-register (MISR) as the ORA. It is possible to use other types of the TPG and the ORA, as well.

There are two operation modes associated with a circuit equipped with BISA, as shown in Table 8.1. In the *normal mode*, the original circuit is working, but the BISA is completely shut down by blocking clocks to LFSR and MISR. BISA stays quiet and does not affect the original circuits. On the other hand, in the *authentication mode*, a slow clock is provided to BISA. In the *test mode*, which is a sub-mode of the authentication mode, LFSR generates N-bit test patterns and applies them to all BISA blocks at every clock cycle. At the same time, each BISA block outputs one bit so that the MISR will receive a total of M bits from M blocks. When a sufficient number of test patterns are applied after a certain number of clock cycles, the BISA circuits can be fully tested. The BISA then stops and the value stored in the MISR is the signature generated from the M BISA blocks' responses. Comparing the obtained signature with the correct signature from simulation shows whether the BISA structure has been tampered with. All registers in LFSR and MISR are connected in a series in the *shift mode*. Therefore, the signature in MISR can be shifted out. In the meantime, the seed for LFSR can be shifted in, as well, during the shift mode.

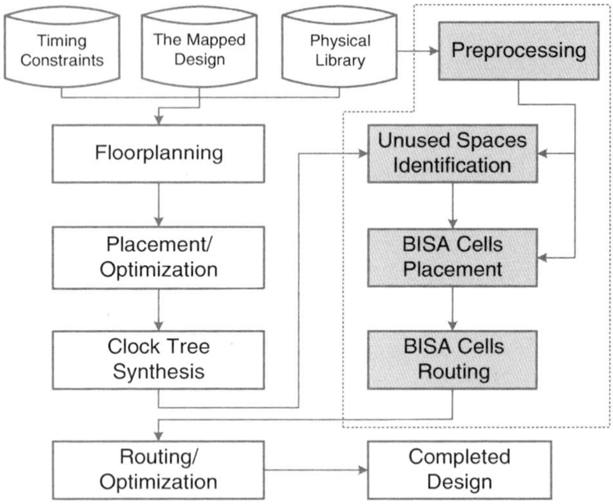

Fig. 8.2 The BISA insertion flow

8.1.2 BISA Insertion Flow

Figure 8.2 shows the general BISA insertion flow. The white rectangles in Fig. 8.2 represent the conventional ASIC design flow. The additional steps required to insert BISA structure are presented in dark rectangles. The BISA insertion flow includes the following steps.

8.1.2.1 Preprocessing

Before starting physical design, some information about standard cells in the technology library should be collected. Firstly, the geometrical information of each standard cell, such as length and width, is required. If the placement direction and location of a particular cell are given, coordinates of its four corners can be calculated. Secondly, the number of input pins and the cell's name are also needed for the routing phase. Thirdly, those cells which will be used as BISA cell should be selected and marked according to following criteria: (1) BISA cells are all minimum-sized cells for different logics so that they are resilient to a resizing attack, which will be explained in Sect. 8.2.2 in detail. (2) The average decoupling capacitance the cell can provide and the average input count should also be considered. Average input count represents the number of inputs of one standard cell compared to the minimum-sized inverter using the same area as the inverter. Similarly, average decoupling capacitance is defined as decoupling capacitance that are standard cell can provide, compared to the minimum DECAP cell in technology library. A large decoupling capacitance from a BISA cell can make up for the absence of DECAP cells [15]. Fewer inputs help improve test coverage, which

Table 8.2 Cell information collected during preprocessing step

Name	Area	Input	Ave. Input	Ave. DECAP.	BISA
DECAP Cell	1,920	0	0	1	No
INVX0	1,920	1	1	0.5	Yes
NAND2X0	1,920	2	2	0.67	Yes
NOR2X0	1,920	2	2	0.67	Yes
INVX1	2,240	1	0.86	1	No
AND2X1	2,560	2	1.5	0.87	Yes
NAND3X0	2,560	3	2.25	0.56	Yes
OR2X1	2,560	2	1.5	0.87	Yes
AND3X1	2,880	3	2	0.83	Yes
NOR3X0	2,880	3	2	0.5	Yes
NAND4X0	2,880	4	2.67	0.53	Yes
NAND2X2	3,200	2	1.2	1.34	No
...					

will be described in Sect. 8.2.1. (3) The smallest cell (usually the inverter) from library must be identified because this method needs the smallest functional cell to ensure that no more cells can be inserted in the unused spaces. Table 8.2 shows an example which is generated with the Synopsys 90nm library [13]. For the purpose of estimating the capacitance, the diffusion capacitance of the contacted source and drain region is comparable to the gate capacitance [12]. In Table 8.2, the columns labeled *Area* and *Input* show each cell's area and input count. Average input count and average decoupling capacitance are obtained, which are listed in the fourth and fifth columns, respectively. For example, NAND2X0 in the third row shows it has twice the input count of an inverter using the same area. The decoupling capacitance ratio of 67% shows that NAND2X0 is 33% less than the DECAP cell in same area. If a cell is selected to be used as a BISA cell, it will be marked in the last column of the table. The cells with the fewest inputs and larger decoupling capacitances are the best candidates to become BISA cells.

8.1.2.2 Identifying Unused Spaces

In order to obtain the location of each cell, physical design tools like the Synopsys IC Compiler [14], can write a DEF file (.def) that contains the coordinates of all placed standard cells. By analyzing the coordinates of all placed cells, the locations of all unused spaces are obtained. All unused spaces are calculated and listed in an Unused Spaces files (.unsp). The format of a UNSP file is shown in Fig. 8.3b. The required information, such as size and location, are listed in the UNSP file.

8.1.2.3 BISA Cells Placement

Designers can obtain the sizes and locations of unused spaces simply by reading the UNSP file. This prevention technique attempts to fill each unused space with BISA

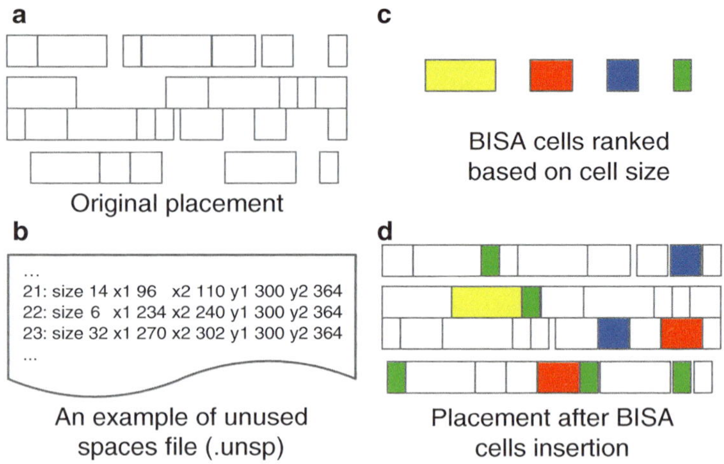

Fig. 8.3 Cells insertion and placement

cells. A greedy algorithm is developed to achieve this goal. First, the BISA cells are sorted by features such as size, average decoupling capacitance, and average input count. Different regions can employ different ranks based on different features. The location and attribute of each region can be given in a constraint file. If one knows a region has high switching probability during simulation, the average decoupling capacitance is more important because BISA cells with larger decoupling capacitors can help reduce voltage drops. Sorting by size and average input pins can achieve a small BISA cell number and high test coverage, respectively. In Fig. 8.3, BISA cells are ranked based on a cell's size. The largest cell (yellow) in Fig. 8.3c has the highest priority for insertion. For each unused space, the largest cell is the first to be tried until the remaining space is smaller than this cell. Then the second largest cell is tried in the same manner. This process will be repeated until the smallest BISA cell has been tried, and eventually no more cells can be added to the layout. The layout after BISA cell insertion is shown in Fig. 8.3d. There are still some white spaces left between cells, but not even the smallest cell (green) can be inserted. The whole process is done by a program which simulates the cell insertion process and produces a tcl command script for physical design tools. Whenever a new BISA cell is inserted, one corresponding command for the physical design tool is generated and saved in the script.

8.1.2.4 BISA Cells Routing

All placed BISA cells need to be connected to form a number of BISA blocks, unlike regular filler cells, which are placed without functionality. The test coverage is a key parameter in connecting BISA cells. First, as many BISA blocks as possible

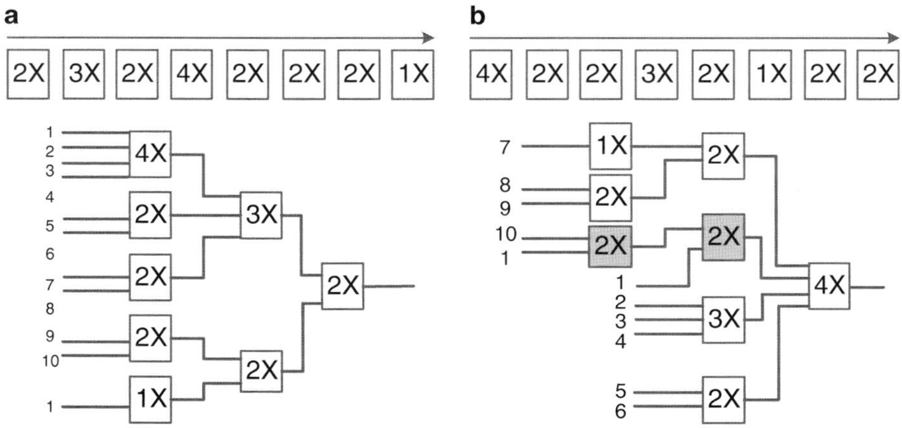

Fig. 8.4 Two examples of BISA cells connection and routing

are created to make each BISA block with fewer gates so that high test coverage is easier to obtain. Before routing BISA cells, one already knows how many BISA blocks will be generated. K BISA-cells are divided into M BISA-blocks equally (M is the size of MISR), so the K/M BISA cells are put in one set. Second, in order to maximize the portion of testable BISA cells, the number of redundant gates in each BISA block should be minimized. To do so, the construction of a tree-structure circuit for each BISA block is proposed. If every input is independent of other inputs, every net becomes controllable and observable in the tree-structure circuit, and the full stuck-at test coverage can be achieved. A tree-structure BISA block is constructed according to the sequence of cells in a block set. Figure 8.4 shows that two different sequences lead to two different tree-structure circuits. The number in each cell indicates its input count. The first gate is used as the root of tree-structure circuitry, which is the top (first) level. Next, x cells (x being the number of inputs of the root cell) are connected to its inputs as its children cells, which are in the second level. The same procedure is repeated for the next levels. Cells are sequentially connected to cells in upper level until all of them are processed, as shown in Fig. 8.4. After completing routing in each block, all the inputs of each block should connect to LFSR sequentially to minimize the number of shared inputs. The outputs from M BISA-blocks connect to a MISR with a size of M.

As cells are added to one block, the number of inputs will increase consistently. If the number of inputs of one block is greater than the number of the LFSR's output, some inputs have to share LFSR outputs in a broadcasting fashion. However, the dependence between several pairs of inputs might result in redundant gates in BISA blocks, thereby affecting fault coverage.

Routing these extra BISA cells will make the routing process more complex. In order to cope with pressure from a limited routing area, the nearest cells are assigned to one BISA block to shorten the interconnections among cells in this block in a greedy fashion. A slow test clock can be used to test BISA in the authentication mode. The slow authentication clock can be estimated according to process technology and the maximum stage of a BISA block. This allows us to ignore the BISA's timing constraints and focus on obtaining maximum coverage and saving routing resources. For example, a BISA can use any available metal layer.

Commands are available to create nets and connect them logically during physical design. Both original design and BISA will be physically routed during routing/optimization. Once the timing and sign-off of the design are successful, the last step involves the generation of the GDSII/OASIS format of the design for final tape-out.

8.1.3 BISA Design in System-On-Chips (SOCs)

SOC is typically a bottom-up hierarchical design, and the top module includes other predesigned sub-modules, or what is known as intellectual property (IP) cores. If all the IP cores are designed with BISA inside by the flow described in Sect. 8.1.2, each IP core can then simply be treated like a regular standard cell, since there is no available space in each IP core after implementing BISA. Therefore, BISA design in the top module of a SOC is almost the same as that in a single module design. One difference is that IP core cells are much larger than normal standard cells. They cannot be placed in a row, so large empty spaces may be left between IP cores. For SOC design, the BISA design flow is modified to fit all unused spaces between IP cores and standard cells.

For a SOC design, organizing LFSRs/MISRs in the top module or sub-modules in order to minimize area overhead requires a specific consideration. Two structure topologies are used, namely distributed and centralized, respectively. A distributed structure means that each IP core and the top module have its own LFSR/MISR. In a centralized structure, one centralized LFSR and one centralized MISR in the top module or one of the sub-modules are used to provide test patterns and compact responses for BISA blocks in both the top module and IP cores.

8.2 Analyzing BISA Structure

8.2.1 BISA Test Coverage

The testable stuck-at-fault test coverage of each tree-structure BISA block theoretically would be 100% if every input is independent. However, if two or more inputs share the same output from LFSR, the dependence among these shared inputs will reduce the controllability/observability and eventually bring down the test coverage. Excessive BISA cells in a BISA block or small LFSR/MISR will result in inputs sharing in this block. As an example, the BISA block in Fig. 8.4b has 11 input pins, but the size of LFSR is 10, which is one bit less than the inputs of the BISA block. Thus, one input pin has to share with another input pin in this BISA block. In Fig. 8.4b, the two dark gates share the first bit of the LFSR (input labeled "1"). The most straight-forward way to ensure full fault coverage is to increase the size of LFSR or MISR.

8.2.2 Potential Attacks

In addition to improving test coverage, BISA should be immune to different attacks that attempt to create space for Trojan gates via removal, redesign, resizing, or bypassing testing. As previously mentioned, BISA cells are the same as other circuit cells, so it is extremely difficult for an adversary to identify them. Assuming this is possible, attacks can be divided into two categories according to their targets, filler cells or original standard cells. Four potential attacks on these two targets are discussed in the following.

8.2.2.1 Removal Attack

A removal attack is the most direct and simplest way to create space for Trojan gates. Simply removing BISA cells will change the functionality of BISA blocks. Test patterns will test the functionality of BISA blocks, and BISA signatures can tell whether the BISA cells are there or not. If some functional standard cells from the original design are removed by adversaries, the original functionality will be altered. Both functional tests and structural tests would be able to detect removal attacks. Using as few unnecessary non-functional standard cells as possible to restrict potential available spaces is recommended.

8.2.2.2 Redesign Attack

An adversary's changes to any gate in the BISA or original circuits are easily detectable because the circuits' functionalities are changed. If an adversary is forced to redesign the layout to achieve the same functionality and insert Trojan gates as well, then the chip dimensions will probably also change. This change could result in a different placement for some or all circuit components. Any changes in the physical layout will likely change the delay and power characteristics of a circuit, which will make it easier to detect the Trojan.

8.2.2.3 Resizing Attack

In all BISA cells are already the minimum size and cannot shrink any further, but some standard cells in original design can be resized to smaller standard cells with the same logic. However, the circuit performance will be affected by using small cells.

8.2.2.4 TPG/ORA Attack

TPG and ORA are used to generate patterns and signatures, so they are potential targets for an adversary. Both TPG and ORA are quite secure against different kinds of attacks. Any modification to register or XOR gates in TPG or ORA will result in totally different patterns and signatures. No matter what change is made, a different signature will indicate that the circuit has been tampered with. If an attacker somehow acquires generated test patterns and the corresponding signature by attacking TPG or ORA, the attacker might then store these results, bypass BISA, and output them, pretending that BISA is working normally. However, even if an adversary obtains all the information, a bypassing attack will be detected, because the seed of LFSR can be changed at any time so that a new set of test vectors and their new signatures will be generated. The adversary cannot predict the seed, therefore, the trustworthiness of BISA can be ensured by applying different seeds.

8.2.3 Yield

The discussions above rely on the fact that BISA is genuine and works without manufacturing defects. However, BISA contains LFSR, MISR, and many BISA cells, and silicon defects are unavoidable during fabrication. A chip producing a faulty signature may contain hardware Trojans, but it also may have been caused by defects in BISA, while the original circuits are working correctly. Of course, good chips should not be discarded for defects that only appear in their BISAs; to do so would reduce yield and increase costs. A hardware Trojan is intentionally inserted into all or a percentage of chips by adversaries, while defects randomly occur due to imperfect manufacturing processes. The probability of two chips with the same defects would be very low, but chips with same Trojan would always produce the same faulty signatures, a distinction that provides opportunities to separate defective chips from Trojan-inserted ones. Note that masks used for fabrication cost millions of dollars, so it might be infeasible for an adversary to make different masks for different chips (or wafers) to imitate random defects. These chips with the same right signatures and same faulty signatures are suspected as infected chips. If the faulty signatures from one chip are completely different from other chips, the faulty signatures most likely have resulted from defects in the BISA circuit.

8.3 Results and Analysis

BISA is implemented in several single module benchmarks (s38417, AES) and SOC benchmarks (System05, vga_lcd, Leon 3 Microprocessor) with the 90nm technology library [13]. Tight timing constraints are used to demonstrate the

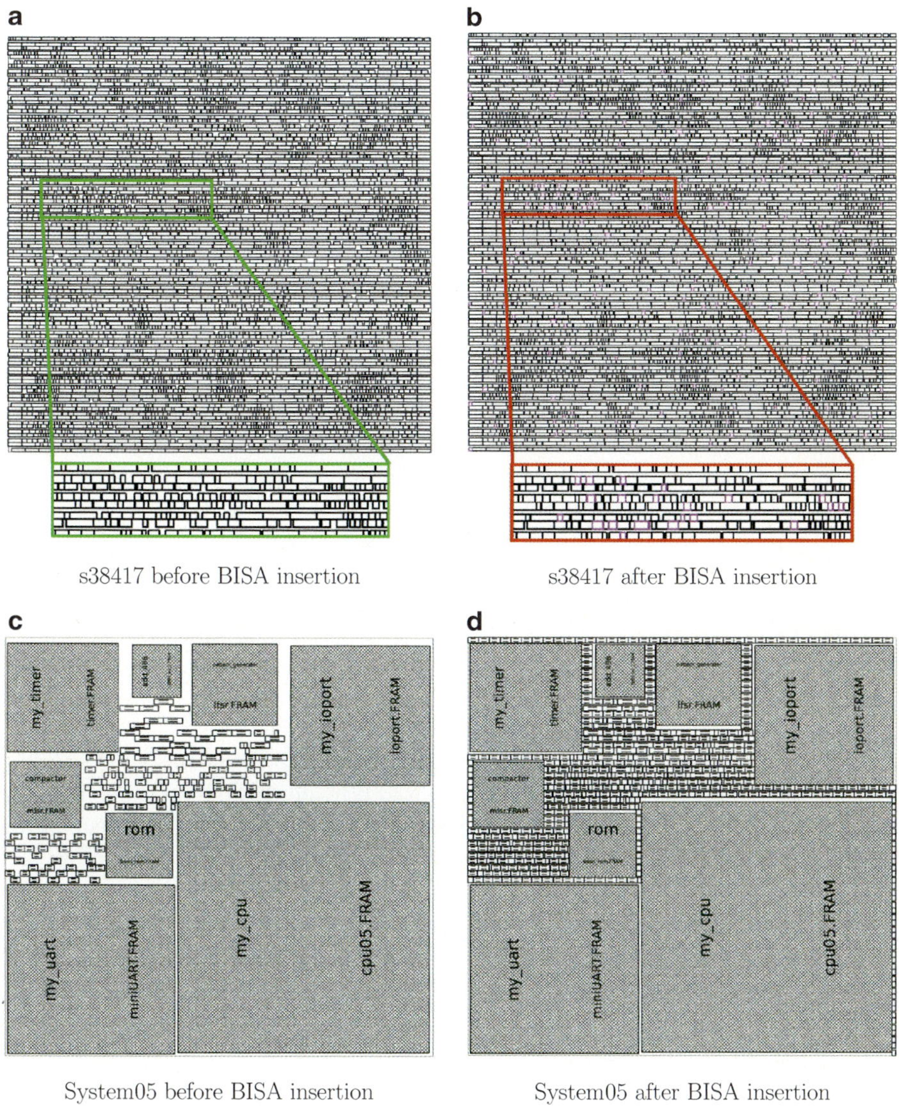

Fig. 8.5 Implementation

feasibility of the proposed technique. Figure 8.5a–d show layouts before and after implementing BISA for s38417 and System05 with the Synopsys IC Compiler [14].

Table 8.3 shows detailed information about the original circuits and BISA circuits in different benchmarks. For some benchmarks several core utilizations (i.e., layout compaction) are investigated. As core utilization increases, fewer BISA cells were inserted into layout to fill the empty spaces. Therefore, high utilization is highly recommended for designs which need BISA.

Table 8.3 BISA implementation under different utilization

Benchmark	Standard cells	Core utilization	BISA cells
s38417	5,242	90%	131
s38417	5,242	80%	180
s38417	5,242	70%	982
AES	26,447	93%	5,925
AES	26,447	90%	5,971
AES	26,447	85%	5,998
System 05	2,284	Fit	418
vga_lcd	124,031	Fit	7,710
Leon 3	545,836	Fit	64,728

In a large circuit, the core utilization cannot be very high due to the use of conservative floorplanning tools. For example, the highest core utilization for the benchmark AES is 93%, while 99% core utilization is still acceptable with a benchmark as small as s38417. Therefore, a lot of white spaces are left in a large circuit, sufficient to place post-design LFSR and MISR. For SOC designs, "Fit" core utilization means the most compact floorplanning that allows all IP cores to be placed without violations. A lot of BISA cells need to be inserted in the three SOC circuits, as shown in Table 8.3. On the other hand, since the gaps between IP cores are very large, large post-design LFSR and MISR can be inserted in these unused spaces to ensure a very high test coverage. BISA does not increase circuit dynamic power. However, it will impact circuit leakage power since BISA components will draw leakage current during functional operation.

Although there is no space for DECAP cells after applying BISA, BISA cells that are selected from standard cells are able to provide decoupling capacitance to some degree [15]. During the BISA preprocessing step, cells with high decoupling capacitors can be used as BISA cells, as described in Sect. 8.1.2.1 and Table 8.2. The BISA cells can still serve as strong sources of decoupling capacitance. The comparison in Table 8.2 shows that BISA cells can provide about two thirds the decoupling capacitance of abtraditional DECAP cell using the same area.

Test coverage is a critical parameter for the BISA technique, as it reflects the confidence level of BISA's result. Figure 8.6 shows the BISA test coverage by performing fault simulation when BISA is inserted into the s38417 benchmark with a core utilization of 80%. In Fig. 8.6a, different sizes of LFSR, such as 32-bit, 16-bit and 8-bit are investigated. After 500 test vectors generated from LFSR are applied, the improvements are 27.63% and 36.1% for 16-bit and 32-bit LFSR, respectively. Since a 32-bit LFSR is able to provide 32 independent inputs to BISA blocks, which is larger than the number of inputs of any BISA block, the test coverage is 100%. Different numbers of test patterns are applied and their coverages are shown in Fig. 8.6b. As the number of test patterns increases, test coverage continuously increases, especially for larger LFSRs. The highest test coverage can be achieved by applying different seeds to generate all required test patterns.

In order to demonstrate that BISA is immune to attacks, ten cases of attacks are made to verify its effectiveness. In the system05 SOC circuit, 418 BISA cells are inserted to fill unused spaces. Size 32 LFSR and MISR are used to form

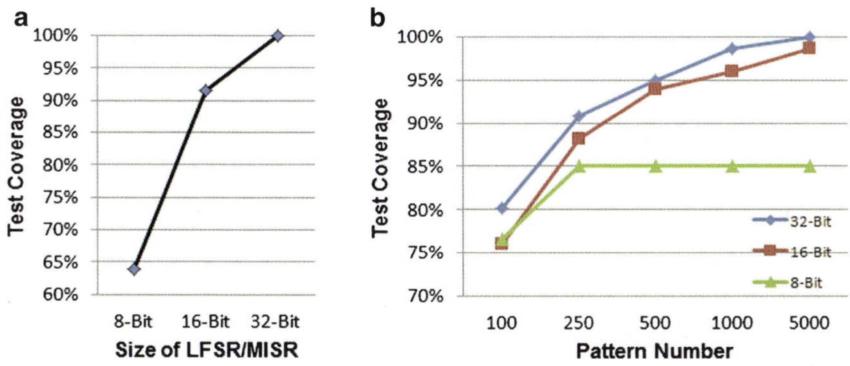

Fig. 8.6 BISA test coverage analysis **(a)** Test coverage improvement **(b)** Test coverage of pseudorandom patterns

Table 8.4 Attack cases

Case	Type	Attack description	Signature
0	Genuine	None	0712022D
1	Removal	Remove a leaf cell	0DA8936E
2	Removal	Remove an internal cell	157F4929
3	Removal	Remove a leaf cell	0ED740FC
4	Removal	Remove an internal cell	D5E2706E
5	Removal	Remove an internal cell	43D51D83
6	Change	Change a leaf cell OR3X1 to AND3X1	F2308684
7	Change	Change an internal cell AOI222X1 to OAI222X1	157F4929
8	Change	Change a root cell AOI222X1 to OAI222X1	F39C3B1E
9	Change	Change a leaf cell AND3X1 to NAND2X1	157F4929
10	Change	Change an internal cell NAND4X1 to NAND3X1	0B17041F

BISA structure. 616 ATPG patterns can reach 99.65% testable coverage. When 500 patterns from LFSR are applied, the stuck-at fault test coverage is 81%. In Table 8.4, case 0 shows the result for the genuine BISA, and the 32-bit signature is shown in hexadecimal format in the last column. Five kinds of gates are selected for separate removal from different BISA blocks. In addition, another five types of gates are selected to be changed to other types of gates in different BISA blocks. The results of these ten cases are shown in Table 8.4. In each case, the signature generated from MISR is different from the genuine signature, which shows that BISA has detected these attacks. In Table 8.4, an internal cell means it has children cells, and a leaf cell is a cell that does not have children cells.

8.4 Summary

The BISA structure is introduced to address IC trust issues attributed to an untrusted foundry. BISA fills all unused spaces to prevent or hamper the Trojan insertion process after completing layout design by leaving no space for Trojan gates. BISA cells are connected to form a certain functionality. BISA has no impact on the original circuit, since BISA and original circuits work independently. Additionally, different kinds of attacks can be detected, ensuring that the BISA's result is trustworthy. By comparing signatures, designers can know whether the chip has been tampered with or not, as demonstrated by the implementation of different attacks.

References

1. M. Tehranipoor and F. Koushanfar, "A Survey of Hardware Trojan Taxonomy and Detection," IEEE Design and Test of Computers, pp. 10–25, January-February 2010.
2. M. Banga and M. Hsiao, "A novel sustained vector technique for the detection of hardware trojans," IEEE VLSI Design, pp. 327–332, Jan. 2009.
3. R. Chakraborty, F. Wolff, S. Paul, C. Papachristou, and S. Bhunia, "MERO: A statistical approach for hardware Trojan detection," Workshop on Cryptographic Hardware and Embedded Systems, pp. 396–410, 2009.
4. D. Agrawal, S. Baktir, and D. Karakoyunlu, P. Rohatgi, and B. Sunar, "Trojan Detection using IC Fingerprinting," IEEE Symp. on Security and Privacy (SP), pp. 296–310, 2007.
5. J.Aarestad, D. Acharyya, R. Rad, and J. Plusquellic, "Detecting Trojans through leakage current analysis multiple supply pad IDDQs," IEEE Transactions on Information Forensics and Security, vol. 5, issue 4, pp. 893–904, 2010.
6. Y. Jin and Y. Makris, "Hardware Trojan Detection using Path Delay Fingerprint," IEEE Int. Workshop on Hardware-Oriented Security and Trust (HOST), pp. 51–57, 2008.
7. H. Salmani, M. Tehranipoor, and J. Plusquellic, "A Novel Technique for Improving Hardware Trojan Detection and Reducing Trojan Activation Time," IEEE Transactions on VLSI, 2011.
8. J. Li and J. Lach, "At-Speed Delay Characterization for IC Authentication and Trojan Horse Detection," IEEE Int. Hardware-Oriented Security and Trust (HOST), pp. 8–14, 2008.
9. Y. Jin, N. Kupp, and Y. Makris, "DFTT: Design for Trojan Test," In Proceedings of the IEEE International Conference on Electronics Circuits and Systems, pp 1175–1178, 2010.
10. R. Chakraborty and S. Bhunia, "Security against Hardware Trojan through a Novel Application of Design Obfuscation," IEEE/ACM Int. Conf. on Computer-Aided Design (ICCAD), pp. 113–116, 2009.
11. http://eda-automation.blogspot.com/2010/06/filler-cells.html.
12. N. Weste and D. Harris, "CMOS VLSI Design: A Circuits and Systems Perspective," 4th edition, Addison-Wesley, 2010.
13. Synopsys Inc., "Synopsys 90nm Generic Library for Teaching IC Design," University program.
14. Synopsys Inc., "IC Compiler, The next-generation physical design system," Synopsys Datasheet, 2010.
15. C. Yeh and M. Marek-Sadowska, "Timing-aware power noise reduction in placement," IEEE Transactions on Computer-Aided Design of Integrated Circuits and Systems, Vol. 26, Issue 3, pp. 527–541, 2007.

Chapter 9
Counterfeit ICs: Taxonomies, Assessment, and Challenges

The problem of counterfeiting of electronic components has drawn a major attention to the media, industry, and government as the counterfeit components market is growing exponentially over the past few years. A study conducted from 2005–2007 [1] reveals that 50 % of original component manufacturers (OCM) and 55 % of distributors (authorized and unauthorized) have encountered counterfeit parts. The most recent data provided by IHS shows that reports of counterfeit parts have quadrupled since 2009 [2]. IHS [3] reports the five most commonly counterfeited components according to the percent of reported counterfeit incidents as: analog ICs (25.2 %), microprocessor ICs (13.4 %), memory ICs (13.1 %), programmable logic ICs (8.3 %), and transistors (7.6 %). Together, these five component groups contribute slightly more than two-thirds of all counterfeit incidents reported in 2011.

The detection of counterfeit electronic components is a multidimensional problem that poses a unique challenge to test engineers. Over the past few years, standards and programs have been put in place throughout the electronics component supply chain that outline the testing, documenting, and reporting procedures [4–6]. However, there is not enough research available to address the detection and avoidance of counterfeit parts. This chapter shall present in detail all types of counterfeits, detection methods, and defects present in the counterfeit parts. Then challenges regarding the implementation of detection methods and their effectiveness are described. To conclude, research opportunities in the domain of counterfeit detection and prevention are explained.

9.1 Counterfeit Taxonomy

The increasing threat of counterfeit electronic components has created a specialized service of testing, detection, and avoidance of such parts. There are different types of counterfeit parts that must be considered during detection, including recycled, remarked, overproduced, defective, cloned, and substandard parts. According to the U.S. Department of Commerce [1], a counterfeit electronic component:

M. Tehranipoor et al., *Integrated Circuit Authentication: Hardware Trojans and Counterfeit Detection*, DOI 10.1007/978-3-319-00816-5_9,

(i) is an unauthorized copy;
(ii) does not conform to the original OCM design, model, and/or performance standards;
(iii) is not produced by the OCM or is produced by unauthorized contractors;
(iv) is an off-specification, defective, or used OCM product sold as new or working;
(v) has incorrect or false markings and/or documentation.

In addition to the above, we also consider overproduced, cloned, and tampered parts as counterfeit electronic parts. Figure 9.1 shows the classification of all different types of electronic components. The counterfeit types can be classified in the seven distinct categories described below.

9.1.1 Recycled

The most widely reported type of counterfeit parts today is the recycled type. It is reported that in today's supply chain, more than 80 % of the counterfeit components are recycled [7]. It was estimated that in 2005, the United States properly recycled only 10–18 % of all electronic waste, though that number has risen to 25 %, as of 2009 [8]. In addition to the US and Europe, so many other developed and developing countries are also improperly disposing of electronics at such an increasingly high rate that they will soon be able to maintain their own recycling facilities. Enforcing recycling in the U.S. alone will not be sufficient to curtail the recycling problem.

It is unlikely that a component can go through a long life in the field and a harsh reclaiming process, and still maintain 100 % functionality. Therefore, parts that are recycled will be of two types, functional but degraded due to aging of their previous usage in the field and non-functional. Aging effects such as negative bias temperature instability (NBTI) and hot-carrier injection (HCI) will create a parametric shift in components; the amount of shift depends on the chip's workload and operation condition. Here, an aged chip is defined as defective or having a measurable parametric shift or degradation in either physical or electrical parameters. A non-functioning part is defined as one that does not perform any of its functions in accordance with OCM specifications.

9.1.2 Remarked

A counterfeit part that is both remarked and recycled has been the most discussed in recent years. The remarking process is described as the removal of the markings on the package (or even on the die) and remarking with forged information. The remarking process is fairly simple. First, a parts' package is sanded or ground down to remove old markings or labels (part number, date code, country of origin, etc.). Then to cover the sanding or grinding marks, a new coating is created and applied

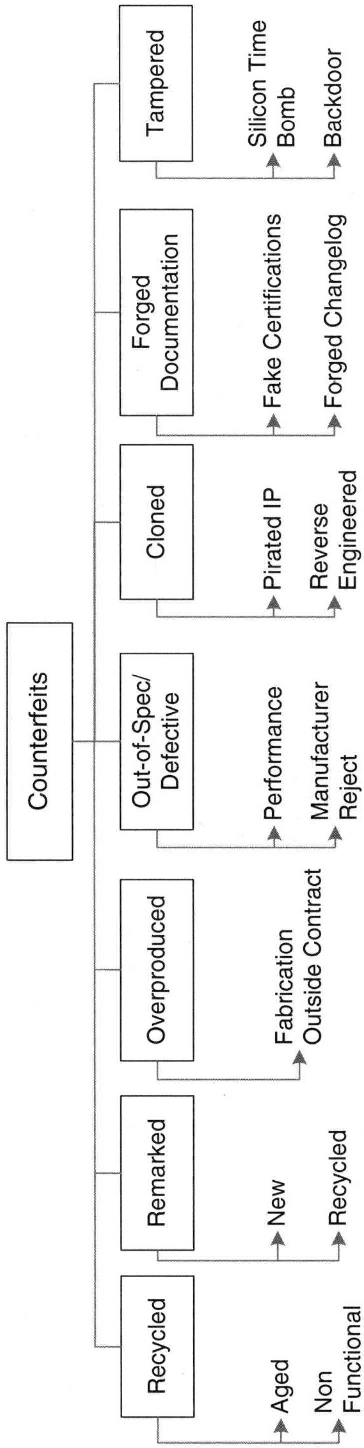

Fig. 9.1 Taxonomy of counterfeit types

to the part. The material used for the coating is either thermal, UV-cured epoxy, or package compatible materials. Once the material is coated, new sets of markings are applied (etched, painted, etc.) on top of the new coating to give the part a proper-looking appearance.

The parts are remarked to obtain a higher or newer specification than what the parts would otherwise rate. For example, a 5-year old part with a 3-year shelf life can be remarked to a more current date to be sold as new, a 1 GHz microprocessor can be marked as 1.2 GHz, or a commercial spec chip can be remarked to a military spec, etc. There are countless ways in which a part can be misrepresented, yet the motivation behind them is the same—to sell at a higher price. It is important to note that remarked parts do not necessarily need to be used prior to the remarking. Either of the previous given examples can apply to new parts as well. It is for this reason that the two sub-categories of remarking are recycled or new parts, as seen in Fig. 9.1.

9.1.3 Overproduced

With the advent of integrated circuits (ICs), electronic components have become more and more complex. As ICs scale in feature sizes and exponentially grow in functionality, they even become more complex. Today's sophisticated ICs are only manufactured in the state-of-art fabrication facilities. Building and maintaining such facilities for the present CMOS technology is reported to be more than several billion dollars altogether and still growing [9]. Given this growing cost and the complexity of foundries and their processes, the semiconductor business model has largely shifted to a contract foundry business model (horizontal business model) over the past two decades. This is also true for the assembly where the dies are packaged, tested, and shipped to the market.

An untrusted foundry/assembly that has access to a designer's IP now has control of chips to fabricate/assemble. Parts strictly in this category go through the exact same processes as authentic and licensed parts. There may not be any danger if a consumer or a supplier ends up with these parts. Instead, the main problem with these parts is that a foundry/assembly is essentially making money without sharing the profit with the design house.

9.1.4 Out-of-Spec/Defective

The other variation of an untrusted foundry sourcing counterfeit parts is an out of specification or a rejected part being sold instead of destroyed. Older parts relied on a basic visual system of inking bad dies. For counterfeits, a simple observation of an ink dot on a die shows that it was never meant to be shipped to market. However, this process has its own set of problems and is being phased out by some manufacturers

[10]. The foundry may either knowingly sell these parts, or they may be stolen and sold in the open market. If these parts make it into the supply chain, authorized distributors without controls could receive counterfeit parts through untested return authorizations. This highlights a critical need for distributor processes that include testing materials being returned from customers. A part which has some parameter out of specification can be dangerous if the designer was relying on that part performing a certain way. On the other hand, the defect may go undetected if the part is never used in that certain way.

9.1.5 Cloned

A cloned part is an unauthorized production of a part without having the legal IP. It is similar to the overproduced counterfeit type. However, there is a clear distinction between the two: whether the counterfeiters have the legal IP or not. Cloning can be done in two ways: they can be reverse engineered or consist of a pirated IP. In reverse engineering, the counterfeiters copy the design exactly and fabricate the part with that design. If a counterfeiter has somehow managed to acquire an IP in an illegal manner without paying the royalty and then fabricates parts using that IP, the parts fall into pirated IP category.

9.1.6 Forged Documentation

This category is quite simply the easiest to fake. If a part is authentic, there is no need for it to ship with forged documentation, so a part that does include forged documentation is also a counterfeit. Forged documentation may include certifications of compliance with some standards or programs, a revision history, or a change-log of a component. A simple analog part will likely not be susceptible to a revision history forgery since its design has been standard for several decades, but a complex VLSI design, which goes through many revisions, is in danger. However, archived documentation for older designs and older parts may not be available at the OCM, making it difficult to verify if the information is authentic. In addition, many organizations have merged or have been acquired over the years resulting in information lost through the transitions.

9.1.7 Tampered

The final category of counterfeit types is the tampered type. Here, we are considering hardware Trojans specifically added to integrated circuits. Such Trojans can either act as a silicon time bomb where the device can behave differently under certain conditions or act as a backdoor where secret information from the chip can be sent out to an adversary. In both cases, the chip behaves outside of its specification, and thus we have included such ICs into counterfeit parts. A detailed taxonomy of such parts can be found in [11].

9.2 Electronic Component Supply Chain Vulnerabilities

Typically, an electronic component will go through a life cycle as shown in Fig. 9.2. The process includes design, fabrication, assembly, distribution, usage in the system, and finally end of life. In this section, the vulnerability of each step to counterfeit emergence is discussed.

The components are stripped off from the printed circuit boards (PCBs), which are recycled back into the supply chain. Recycling brings components that were at their end-of-life phase and puts them back into distribution. The counterfeiters remark these parts and sell them as new in the open market. Also, the new parts can be remarked to upgrade the specification (e.g., upgrade to industrial/space grade from commercial grade components) mainly due to profitabilty of those higher grade components. Second, parts that have been overproduced are introduced into the market from a chip fabricator. The overproduction of parts happens when one of the entities or rogue employees involved in fabrication process produces more than they were contracted to produce, as discussed earlier. These extra parts are taken from the foundry and sold independently on the open market or on internet marketplaces. Third, cloned parts go through a similar process to that of an authentic part, but because they are not authentic they are not subject to the same processes, tests, and standards as the end user was anticipating. This poses the problem that not all defects will be detected during the limited testing time. There is no guarantee that the part will function as planned. Fourth, defective and out-of-spec parts can

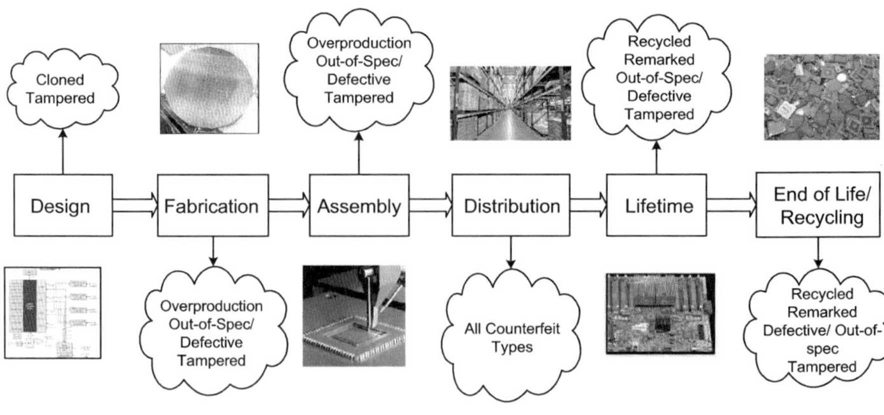

Fig. 9.2 Supply chain vulnerabilities for electronic components

get into the supply chain at various steps, such as design, assembly, distribution, lifetime, and end of life. Finally, tampered parts can potentially enter into supply chain at any part of the life cycle.

9.3 Counterfeit Defects and Detection Methods

Counterfeiting is a multidimensional problem that requires multiple test methods to detect the various anomalies introduced due to the threat. Various types of counterfeits (described in Sect. 9.1) directly impact the detection methods. The detection of remarked counterfeit components is different than the cloned or overproduced types. The types of defects present in remarked type are related to the remarking process (remarking on the package or die) and the change of electrical parameters when the component was used before. The type of anomalies related to the overproduced or cloned types are detected based on the manufacturing process. Due to the complexities and variations of counterfeit methods, a comprehensive taxonomy of defects existing in all counterfeit types in presented in the following. The classification of test methods is also very important as each method can detect a group of defects. Note that here defects are defined as any anomalies and changes seen in parts that are different than the authentic ones. A defect could as simple as an invalid date/code or as complex as some of the hard-to-detect electrical defects.

9.3.1 Defect Taxonomy

Figure 9.3 presents the classification of the defects present in the counterfeit components. The defects are broadly classified into two major categories, i.e., physical and electrical defects. Tampered defects are not discussed in the taxonomy, as additional research for detection is needed in this area.

9.3.1.1 Physical Defects

Physical defects are directly related to the physical properties of the components. They can be classified as exterior and interior defects, depending on the location of the defect related to the packaging. Exterior defects are:

(i) *Packaging/Shipping*: The most obvious defects will be ones that are associated with the packaging or shipping of the parts arrived in. There are some common defects, namely invalid lot code, invalid OEM shipping and packaging labels, missing moisture sensitive device (MSD) indicators, etc.

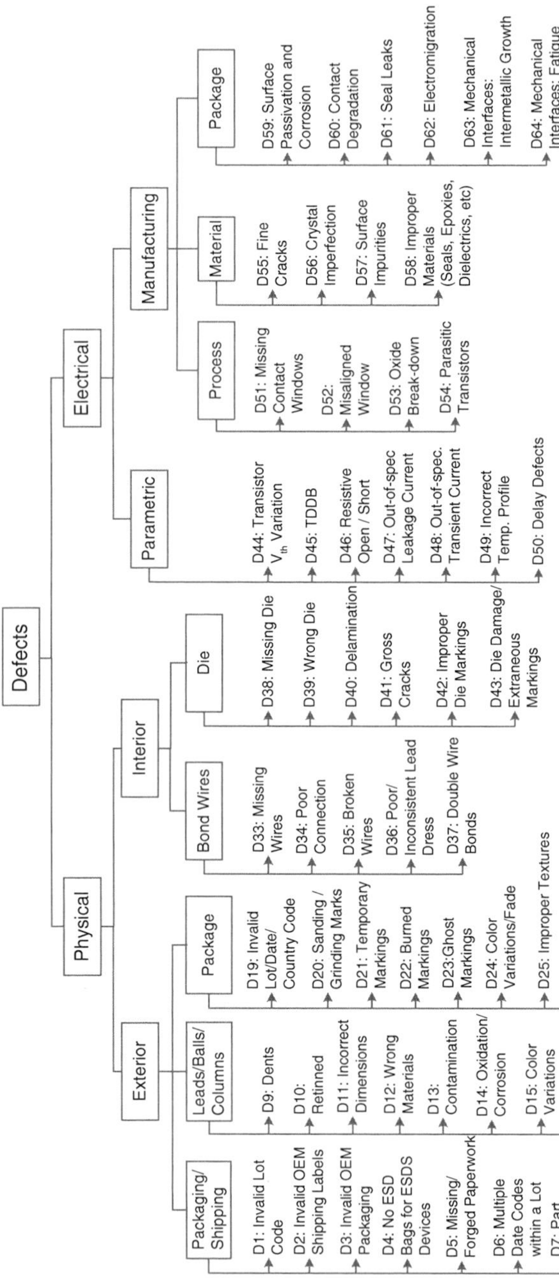

Fig. 9.3 Taxonomy of defects presented in counterfeit components

(ii) *Leads/Balls/Columns*: The leads/balls/columns on an IC can show how the part has been handled if it was previously used. Physically, they should adhere to datasheet specifications, including size and shape. The final coating on the leads should conform to specification sheet.
(iii) *Package*: The package of an IC can reveal significant information about the chip. As this is the location where all model numbers, country of origin, date codes, and other information are etched, counterfeiters are usually careful not to damage anything while keeping the package looking as authentic as possible.

Interior defects are located inside the package. They are mainly divided into two types. They can be either bond wire or die-related defects.

(i) *Bond Wires*: Some common defects related to bond wires are missing bond wires inside the package, poor connection between the die and bond wire, etc.
(ii) *Die*: The die reveals a lots of relevant information regarding the component. The defects present in the die are from die markings, cracks, etc.

9.3.1.2 Electrical Defects

Typical electrical defects can be classified into two distinct categories. They are parametric defects and manufacturing defects. The main reason for adding manufacturing defects under the electrical category is that these defects can be almost completely detected by manufacturing tests (a.k.a, electrical tests, will be discussed in the following section).

(i) *Parametric Defects*: Parametric defects are the manifestation of the shift of component parameters due to prior usage or temperature. A shift in circuit parameters due to aging will occur when a chip is used in the field for some time.
(ii) *Manufacturing Defects*: The defects under this category come from the manufacturing process. These defects are classified into three categories—process, material, and package.

9.3.2 Detection Method Taxonomy

Figure 9.4 shows a taxonomy of counterfeit detection methods arranged the same way the defects are, with the relative ease of test from left to right. The boxed entries represent what will be called a class of tests. Each class contains either similar tests, or will detect similar defects. The objective of developing such a taxonomy is to identify all potential test techniques to help detect the following counterfeit types: recycled, remarked, overproduced, cloned, defective, and out-of-spec. The last two types, forged documentation and tampered, are not considered in this taxonomy.

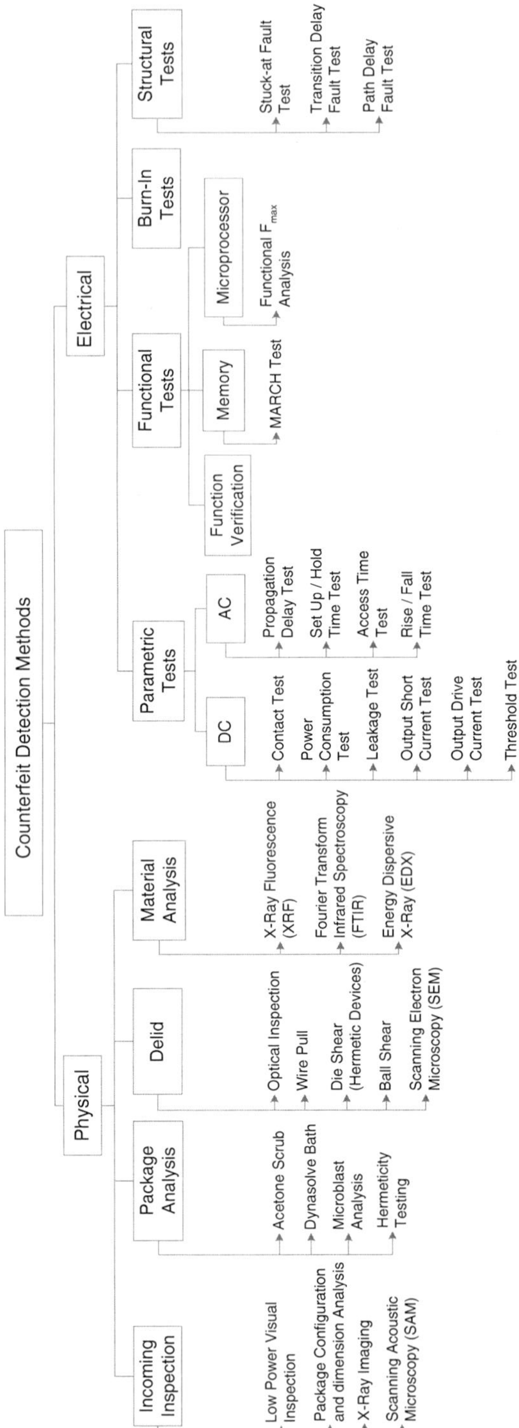

Fig. 9.4 Taxonomy of counterfeit detection methods

9.3.2.1 Physical Methods

Physical methods are classified into four major categories:

(i) *Incoming Inspection*: When an order is first received, it goes through the incoming inspection. All parts are inspected thoroughly in this step. Incoming inspection is divided into three categories: low-power visual inspection, package configuration and dimension analysis, and X-Ray imaging. All parts should be strictly documented and inspected during incoming inspection. The physical dimensions of the component are measured either by hand-held or automated test equipment in the package configuration analysis. In X-Ray imaging, the internal structures of the components are observed in real time.

(ii) *Package Analysis*: Components that appeared suspicious during incoming inspection should be used as subjects in this class of tests. Blacktop testing is the procedure of testing a part's resistance to various solvents. A non-epoxy blacktop coating should be resolved in acetone, while a thermal or UV-cured epoxy will require the use of a much more aggressive solvent. Microblasting analysis is a dry and superfine blasting process. Various blasting agents with proper grain sizes are bombarded on the surface (package) of the component, and the materials are collected for analysis. Microblasting analysis may require much more sophisticated test techniques, such as Laser Scanning Microscoppy, to detect anomalies on package types that have been microblasted with media that do not embed in the package.

(iii) *Delid*: Delid is a process by which the inspection of the internal structure, the top surface of a die, bond wires, or metallization traces, etc., of an electronic component can be performed.

(iv) *Material Analysis*: The chemical composition of the component are verified using material analysis. There are several tests that can perform material analysis including X-ray fluorescence (XRF), Fourier transform infrared (FTIR), energy dispersive X-ray spectroscopy (EDX), etc.

9.3.2.2 Electrical Methods

In this section, we will discuss various manufacturing and process-related tests suitable for detecting a counterfeit chip. An automatic test equipment (ATE) [12] may be required for some of the tests.

(i) *Parametric Tests*: Parametric tests are performed to measure the parameters of a chip. Defects can be observed at any of the chip's input/output (I/O) pins. A defect will modify the observed voltages/currents/delays at these I/O pins, which can be detected using parametric tests. There are several parametric tests currently used as manufacturing tests, and they are contact test, power consumption test, output short current test, output drive current test, threshold test, rise and fall time tests, set-up, hold and release times tests, propagation delay tests, etc.

(ii) *Functional Tests*: Functional tests are the most efficient, and perhaps most expensive, way of verifying the functionality of a component. Read/write operations are performed on a memory to verify its functionality. For example, for a memory chip, MARCH tests can be applied for counterfeit detection. Functional f_{max} analysis can be applied to detect counterfeit microprocessors.
(iii) *Burn-In Tests*: The device is operated at an elevated temperature to simulate a stress condition to find infant mortality failures and unexpected failures to assure reliability.
(iv) *Structural Tests*: Structural tests are very effective for detecting the manufacturing defects for out-of-spec/defective counterfeit types. It can detect the cloned (reversed engineered) counterfeit components if there are some anomalies in the reverse engineering process. It can also detect some of the delay defects due to aging in recycled and remarked counterfeit types. Its effectiveness is questionable, however, for obsolete parts where structural test programs may not be available.

9.4 Challenges Ahead and Roadmap

There are major challenges that must be overcome for the development of effective test methods (discussed in Sect. 9.3.2). Every attempt should be made to stay ahead of counterfeiters to prevent a widespread infiltration of such parts into our critical infrastructures.

9.4.1 Current State of Knowledge

Today's electronic supply chain is polluted by counterfeit components. Not only obsolete parts, but also active parts pose a great threat to the supply chain. Obsolete parts are defined as parts that are no longer manufactured by the OCM, while active parts are those parts that are currently being fabricated because of market demand, but the OCM does not change their chip design or fabrication process. In both cases, there is no opportunity to design the chip with counterfeiting in mind.

There are several standards (such as AS6171 under development by the SAE G-19A committee, CTI CCAP-101, and IDEA STD-1010) in place to guide the user for counterfeit detection and avoidance. However, all these standards try to mainly deal with two types of counterfeits—recycled and remarked. In addition, these attempts focus on existing test techniques which have proven to be inefficient on tampered chips and could be ineffective as counterfeiters improve their skills and techniques. Further, there are currently no simple methods to verify components as genuine if they belong to overproduced and cloned ICs. There are no metrics to be able to effectively evaluate the efficiency of the proposed techniques. Another problem is the consistency of analysis from different test laboratories. Many of the techniques are qualitative and require significant experience to conduct testing effectively. Very often a chip is marked as authentic by one test lab and marked as counterfeit by another test lab. Such inconsistency in the detection of counterfeit parts can have catastrophic effects if the chip is deployed into a critical application.

9.4.2 Detection and Prevention Policies

Since 2011, counterfeit electronics have gained significant attention from policy makers because of the prevalence of such parts in the electronics component supply chain and their implication on critical applications, particularly in the defense industry. The U.S. Congress enacted the National Defense Authorization Act (NDAA) of 2012 [13] on December 2011. Section 818 of NDAA 2012 contains new requirements for the Department of Defense (DoD) to detect and avoid counterfeit electronic parts. As a part of the assessment process, the Secretary of Defense shall review the DoD acquisition policies and systems for the detection and avoidance of counterfeit parts. These regulations have been imposed on contractors and trusted suppliers to avoid the entry of counterfeit parts in the supply chain. The detection of a suspect part must be reported within 60 days to the appropriate Government authorities and the Government-Industry Data Exchange Program (GIDEP). A risk-based methodology shall be implemented by the Secretary of Homeland Security for products purchased by DoD. Contractors must address the proper training of personnel, the traceability of parts and the proper inspection, and the testing and reporting of parts. Severe offenses and penalties are imposed for trafficking counterfeit parts and services. We believe such policies must be updated frequently, as counterfeiters are advancing in the way they get their parts into electronics systems, especially as more systems are being integrated off-shore and deployed to the United States.

9.4.3 The Need for Evaluating the Effectiveness of Detection Methods

Detection methods are not uniform to all components. A set of tests can be useful to detect a specific type of part (e.g., microprocessors, memories, etc.) but may not extend to other part types (analog, passive parts, etc.). Physical methods can be applied to all part types, however, some of the methods are destructive and take hours to test. As a result, sampling is done to certify a batch of parts by observing a small number of parts. On the other hand, conventional electrical test methods are non-destructive and time efficient, yet they can be very expensive because such techniques are not necessarily designed for counterfeit detection. In addition, fixtures may need to be produced and code may need to be written to test various parametric values. Testing according to the original manufacturer specifications may not be possible for complex chips in the absence of the original manufacturer test tapes and fixtures. Electrical test techniques are advantageous because no sampling is required and all parts can be tested. There are some issues associated with electrical tests that must be addressed, as well.

9.4.3.1 Physical Methods

There is a distinct challenge in implementing physical detection methods uniformly among the components.

(i) *Sampling*: Most physical tests are destructive. Sample preparation is extremely important as it directly relates to test confidence. If a few counterfeit components are mixed into a large batch, the probability of selecting the counterfeit one is extremely small.
(ii) *Test time and cost*: The test time and cost are major limiting factors to implement physical tests for counterfeit detection. The equipment used for the physical inspection of such parts (e.g. scanning electron and acoustic microscopy (SEM or SAM)) are not custom designed for detecting counterfeit parts. The test time for performing some tests even on a single component is of several hours.
(iii) *Automation*: The tests are done in an ad-hoc fashion with no metrics for quantifying against a set of counterfeit types, anomalies, and defects. Most of the tests are carried out without automation. The test results mostly depend on subject matter experts (SMEs). The decision-making process is entirely dependent on the operator (or SMEs). This is indeed error prone. A chip can be considered counterfeit in one lab while it could be marked as authentic in another lab. This was proven by a test run by G-19A group that some labs reported the chip as counterfeit and some others as authentic [14].

9.4.3.2 Electrical Methods

Electrical tests have the potential to be the only efficient tests that can be implemented for counterfeit detection without sampling. However, there are major challenges to implementing such tests. One limitation is that recycled chips may perform similarly to new chips but may demonstrate latent defects and have reliability concerns. The electrical tests described in Sect. 9.3.2.2 have their own limitations.

(i) *Parametric Tests*: Parametric tests are generally very time efficient, taking only a few seconds to run, depending the type and size of the chip. However, due to increased process variation (10–15 % process variation [15]) and environmental variations (temperature, noise, aging, etc.), the electrical parameters of a component vary significantly. It will be very difficult to conclude if the variations in the parameters of a component are due to aging (for recycled and remarked components) or to process variations in the circuit. One can perform a statistical analysis based on the data observed from the parametric tests to determine the confidence level that a part is counterfeit with or without a golden IC.

(ii) *Functional Tests*: There are major limitations to implementing these tests in counterfeit detection. For an analog circuit, it is easier to test the functionality of the chip. However, it is extremely difficult to verify the functionality of modern digital VLSI chips with millions or billions of transistors. The major issues associated with functional tests for counterfeit detection include: (a) difficulty of generating effective test patterns targeting all nodes in the circuit under test, (b) increased test cost, as it may contribute up to 30–40 % of the total production test cost. It may not be as much of a concern from the counterfeit detection perspective as there are many fewer testable components than new manufactured components. However, test program generation for obsolete and active parts where only limited knowledge of the part is available will be extremely difficult, if not impossible. (c) the requirement for a high-speed tester in order to apply functional test patterns to DUT, and finally, (d) it is fairly impossible to get the complete set of test vectors for an obsolete part from the OCM. In some cases, the OCM may no longer exist or the required information may no longer be available in archived records.

(iii) *Burn-In*: While burn-in testing is useful in detecting the failures of components with defects, the time and cost it takes to develop functional test vectors to perform a burn-in test makes it attractive and useful only for high-risk applications.

(iv) *Structural Tests*: The implementation of structural tests in counterfeit detection is extremely challenging because of the following reasons: (a) The structural tests require total access to the internal scan chains of a component. Sometimes, the IP owners do not give others permission to access their design and disable the internal scan chains with a fuse. (b) Obsolete parts may not have design for testability (DFT) structures implemented. (c) If there is no netlist present for a component, it is not possible for ATPG to generate test vectors for testing those components.

9.4.4 Roadmap and Research Opportunities

Three distinct types of components are making the electronic component supply chain vulnerable, namely obsolete, active, and new components. These different types of components pose different threats to the entire supply chain. Obsolete parts are no longer manufactured, and active parts are being fabricated based on the previous design and developed masks. Therefore, the focus should be on the detection of counterfeit components first to avoid installing a counterfeit part on a circuit card assembly, and then on preventive actions that can be deployed on newly fabricated chips going forward. As previously described, all the standards are in place for the detection. However, there is not enough research done for the prevention of these parts getting into the supply chain as counterfeit parts. Currently, DoD mandates [16] DNA marking [17] to be placed on the chip to tag the components throughout the supply chain. However, the effectiveness of determining authenticity through tagging components by applying DNA marking has yet to prove efficient. DNA tagging should be useful in determining a point of origin when the parts were tagged, but it will not authenticate legacy material or prevent tampering from intermediaries in the supply chain. Additional research and low-cost prevention methods are needed for obsolete parts to ensure, after they are tested and verified, they will not be back in the market as a counterfeit part.

Dealing with new designs could be easier since new mechanisms can be put in place during chip design so that if the chip is subjected to counterfeiting, it can be detected much more easily. The first set of such methods for preventing and detecting new counterfeits designed and fabricated out of recycled materials have been presented in [18–20]. The technique in [18] inserts a light-weight sensor into the chip to capture the usage of the chip in the field and provides extremely easy detection capability. This type of sensor relies on the aging effects of MOSFETs to change a ring oscillator frequency in comparison with the golden one embedded in the chip. As a part used in the field ages because of the wearout mechanisms such as NBTI and HCI, the shift in the frequency of this sensor indicates the level of aging, and provides a simple readout of the value. But more research is needed to provide greater prevention capability against counterfeiting in the first place.

Care must be taken to monitor counterfeiters' activity over time. As counterfeiters become more and more advanced, existing techniques will probably be defeated. More novel techniques must emerge which are dynamic and adaptable to the trends of counterfeit parts. Active and obsolete parts must be monitored and studied for the subtle changes/anomalies to ensure that the part is a new type of counterfeit that has not been documented before. Sharing information between various entities such as the government, industry, test labs, and academia is very important to monitoring counterfeit activity. When information on a new counterfeit type is available, that should be shared to update existing knowledge. The problem of counterfeiting is not going to disappear, but rather will grow more and more and will become sophisticated overtime because of the high profit the counterfeiters gain. Cooperation between government entities, private industries, academia, and test labs across the board will make us all more prepared for how to deal with both current and future types of counterfeit parts. Research in academia in this domain has been almost non-existent. It is important for government and funding agencies to provide more resources and support to academic researchers to help develop innovative solutions for this very challenging problem.

9.5 Summary

There are variety of counterfeit parts currently corrupting the electronic supply chain and current efforts to address the counterfeiting problem are not sufficient. Currently, there is not enough research carried out in the field of counterfeit detection and avoidance in academia and industry. More research is needed to implement effective test methods that are adaptable, as the counterfeiting process will become more sophisticated over time. Finally, all entities involved in the supply chain should collaborate to ensure that no suspect parts finds their way into critical infrastructures.

References

1. U.S. Department Of Commerce, "Defense Industrial Base Assessment: Counterfeit Electronics," Jan. 2010.
2. IHS, "Reports of Counterfeit Parts Quadruple Since 2009, Challenging US Defence Industry and National Security," Apr. 2012, http://www.ihs.com/images/IHS-iSuppli-Reports-Counterfeit-Parts-Quadruple-Since-2009.pdf.
3. IHS, "Top 5 Most Counterfeited Parts Represent a $169 Billion Potential Challenge for Global Semiconductor Market," 2011, http://press.ihs.com/press-release/design-supply-chain/top-5-most-counterfeited-parts-represent-169-billion-potential-cha.
4. SAE, "Counterfeit electronic parts; avoidance, detection, mitigation, and disposition," 2009, http://standards.sae.org/as5553/.
5. CTI, "Certification for coutnerfeit components avoidance program," Sept. 2011, http://www.cti-us.com/pdf/CCAP101Certification.pdf.
6. IDEA, "Acceptability of electronic components distributed in the open market," http://www.idofea.org/products/118-idea-std-1010b.
7. L. W. Kessler and T. Sharpe, "Faked Parts Detection," 2010, http://www.circuitsassembly.com/cms/component/content/article/159/9937-smt.
8. U.S. Environmental Protection Agency, "Electronic waste management in the united states through 2009," May 2011.
9. Mouli, C. and Carriker, W., "Future Fab: How software is helping Intel go nano–and beyond," *IEEE Spectrum*, vol. 44, no. 3, pp. 38–43, 2007.
10. M. Banke, "Implementing inkless wafer sort," July 2006.
11. M. Tehranipoor and F. Koushanfar, "A survey of hardware trojan taxonomy and detection," *IEEE Des. Test*, vol. 27, no. 1, pp. 10–25, 2010.
12. A. Grochowski, D. Bhattacharya, T. Viswanathan, and K. Laker, "Integrated circuit testing for quality assurance in manufacturing: history, current status, and future trends," *Circuits and Systems II: Analog and Digital Signal Processing, IEEE Transactions on*, vol. 44, no. 8, pp. 610–633, Aug. 1997.
13. US Congress, "National Defense Authorization Act for Fiscal Year 2012." [Online]. Available: http://www.gpo.gov/fdsys/pkg/BILLS-112hr1540enr/pdf/BILLS-112hr1540enr.pdf
14. CHASE, 2013, http://www.chase.uconn.edu/arochase-special-workshop-on-counterfeit-electronics.php.
15. ITRS, "International technology roadmap for semiconductors," Nov. 2006, http://public.itrs.net/.
16. U.S. Defense Logistics Agency, "DNA AUTHENTICATION MARKING ON ITEMS IN FSC 5962," Aug. 2012. [Online]. Available: https://www.dibbs.bsm.dla.mil/notices/msgdspl.aspx?msgid=685

17. M. Miller, J. Meraglia, and J. Hayward, "Traceability in the age of globalization: A proposal for a marking protocol to assure authenticity of electronic parts," in *SAE Aerospace Electronics and Avionics Systems Conference*, Oct. 2012.
18. X. Zhang, N. Tuzzio, and M. Tehranipoor, "Identification of recovered ics using fingerprints from a light-weight on-chip sensor," in *Proc. of IEEE on Design Automation Conference*, June 2012, pp. 703–708.
19. X. Zhang, K. Xiao, and M. Tehranipoor, "Path-delay fingerprinting for identification of recovered ics," in *Proc. of IEEE International Symposium on Defect and Fault Tolerance in VLSI and Nanotechnology Systems*, Oct. 2012.
20. J. Villasenor and M. Tehranipoor, "Are you sure it's new? the hidden dangers of recycled electronics components," in *to appear in IEEE Spectrum*, 2012.

Chapter 10
Counterfeit ICs: Detection and Prevention of Recycled ICs Using On-Chip Sensors

As discussed in previous chapter, the counterfeiting of ICs is a major issue impacting the security of a wide variety of electronic systems. A counterfeit component is defined as an electronic part that is not genuine because it [1]: (i) is an unauthorized copy; (ii) does not conform to the original component manufacturer's design, model, and/or performance; (iii) is not produced by the original component manufacturer or is produced by unauthorized contractors; (iv) is an off-specification, defective, or used original component manufacturer's product sold as "new" or working; (v) has incorrect or false markings and/or documentation.

The Office of Technology Evaluation, part of the U.S. Department of Commerce, reported over 10,000 incidents involving the resale of used or defective ICs from 2005 to 2008 alone, much more than other types of counterfeits [1] (shown in Fig. 10.1). As the figure shows, the number of reported incidents of used ICs being sold as new or remarked with a higher grade is much more than other types of counterfeits. In 2008, Business Week published an investigation that traced recycled ICs found in U.S. military supplies back to their sources [2]. It is reported in [3] that used or defective products account for 80–90 % of all counterfeits sold worldwide. With such an estimate, and with the numbers relating to semiconductor sales and counterfeiting in general (presented in [4]), it is possible that the intentional sale of used or defective chips in the semiconductor market could account for about $15 billion USD of all semiconductor sales in 2008 alone. Note that this number may actually be much larger since many of the counterfeit ICs go undetected and are being used in systems today. In addition, the trends cited in [1], suggest that this number will only increase over time.

These used or defective ICs enter the market when electronic "recyclers" divert scrapped circuit boards away from their designated place of disposal for the purposes of removing and reselling the ICs on those boards. The detailed recycling process is shown in Fig. 10.2. After a careful cleaning, those used ICs look like new and can be reused in critical applications. It is vital to prevent recycled ICs from entering critical infrastructures such as aerospace, medical, and defense supply chains since they will fail sooner and less predictably than new chips.

M. Tehranipoor et al., *Integrated Circuit Authentication: Hardware Trojans and Counterfeit Detection*, DOI 10.1007/978-3-319-00816-5_10,

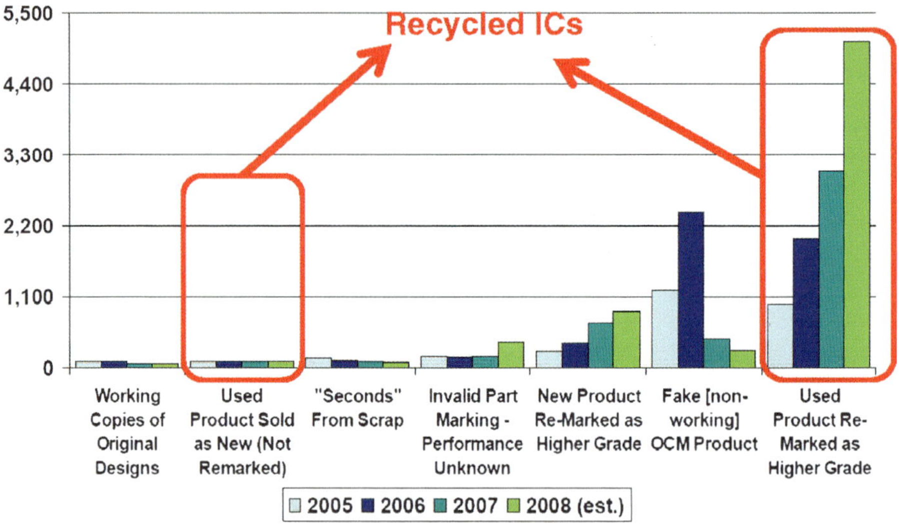

Fig. 10.1 Counterfeit incidents by type of problem for microcircuits from 2005 to 2008

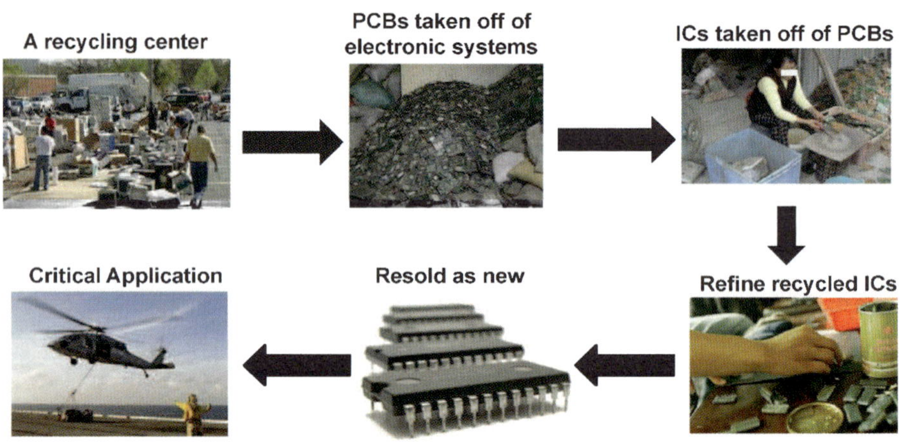

Fig. 10.2 ICs recycling process

Since the recycling process usually involves a high-temperature environment to remove ICs from their boards, there are several security issues associated with these ICs: (i) a used IC can act as a ticking time bomb [5] since it does not meet the specification of the unused (new) ICs; (ii) an adversary can include additional die on top of the recycled die, carrying a back-door attack, sabotaging circuit functionality under certain conditions, or causing denial of service [6]. Therefore, it is vital to prevent these recycled ICs from entering critical infrastructures.

Note that, the term *recycled IC* is used in this book to denote used ICs being sold as new or remarked with higher grades. The terms *unused IC* and *new IC* represent the ICs that are brand new. On the other hand, most of ICs used in the field are not turned on all the time. Take an IC used in a cell phone for example. The cell phone may only be powered on during the day. The real (power-on) usage time of the IC would be much shorter than the usage time with power-off intervals. The term *usage time* is used to represent the accumulated power-on time, even if the IC was used intermittently.

In general, recycled ICs have the same appearance, functionality, and markings as the devices they are meant to mimic, but they have been used for a period of time before they were resold. Even the best visual inspection techniques will have difficulty identifying these ICs with certainty [7]. Physical tests, described in [8], are often used to identify recycled ICs through visual inspection, blacktop testing, scanning electron microscopy (SEM), scanning acoustic microscopy (SAM), x-ray imaging, x-ray fluorescence, fourier transform infrared (FTIR) spectroscopy, etc. These methods can efficiently detect recycled ICs with mechanical defects, such as defects in packaging, lead, bond wires, die, etc. However, they cannot detect recycled ICs without these physical defects. Moreover, physical test cannot be verified all the ICs in a batch of chips as most of these tests are based on sampling. On the other hand, electrical detection methods can be applied to all the ICs under test. SAE AS5553 [8] incorporates some electrical tests, such as the DC curve trace, full DC test, key (AC, switching, functional), and full functional tests at ambient temperature and over temperature in their detection procedures. However, the applicability of these tests to today's complex ICs (microprocessors, memories, programmable logic devices, ASICs, etc.) are major concerns. Using electrical tests for IC detection has not yet been verified completely, and there are currently no available documents to guide recycled IC detection using electrical tests. Therefore, new techniques need to be developed to help measure recycled ICs' specifications and effectively detect them if they have already been used in the field, even for a short period of time.

Over the past several years, several techniques related to recycled IC detection have been developed. Physical unclonable functions (PUFs) implement challenge and response authentication for IC identification [9–13]. For each physical stimulus, the circuit will react in an unpredictable way due to the complex interaction of the stimulus with the physical structure of the PUF and the inherent process variations. As the physical variations for each IC are unique, a distinct ID can be obtained for each IC through the PUF. Techniques to protect ICs against counterfeiting via active and passive authentication and identification (also known as hardware metering) have been proposed in [14–16]. Metering techniques attempt to ensure that the overproduction of ICs will be prohibited. The above approaches are effective at authenticating ICs but not at identifying recycled ICs since they are expected to have the same IDs as the unused ICs.

The computer-aided design and reliability research community has also seen extensive research on the aging of ICs. In particular, ring oscillator based reliability analysis has become a common practice. For instance, a silicon odometer has been proposed to monitor different types of aging effects [17, 18]; however, the objective was to improve the reliability of ICs, not to identify the recycled ICs. If such sensors are used to detect recycled ICs, they will be ineffective because of the presence of process and environmental variations.

The first attempt to identify recycled ICs is presented in [19], a technique that used on-chip sensors. The simulation and silicon results demonstrated the effectiveness of the approach. This chapter will focus on this technique for recycled IC identification.

The major difference between recycled ICs and unused ICs is that recycled ICs have already been used and have experienced aging, as they were removed from their original boards and re-sold in the market. Aging effects, such as negative bias temperature instability (NBTI) and hot carrier injection (HCI), would have had an impact on recycled ICs' performance due to the change in threshold voltage. In this section, two techniques using lightweight sensors (RO-based and AF-based) will be presented to help with the detection of recycled ICs [19].

The RO-based sensor is composed of a reference ring oscillator (Reference RO) and a stressed ring oscillator (Stressed RO). The Stressed RO is designed to age at a very high rate by using high threshold voltage gates to expedite aging so that ICs used for a period of time can be identified. The Reference RO is gated off from the power supply during chip operation, so that it experiences less stress. The frequency difference between the two ROs could denote the usage time of the chip under test (CUT); the larger the difference is, the longer the CUT has been used, and with a higher probability the CUT could be a recycled IC. The impact of intra-die process variations can be minimized with the close placement of the two ROs in the RO-based sensor. Data analysis can effectively distinguish the frequency differences caused by aging from those caused by temperature and inter-die process variations to identify recycled ICs, which is demonstrated by the simulation and silicon results. The RO-based sensor presents a negligible area overhead, imposes no constraint on circuit layout, and is resilient to removal and tampering attacks. The three working modes of the RO-based sensor ensure that the Reference RO cannot be gated on alone, thus the frequency difference between the two ring oscillators cannot be changed to mask detection.

The AF-based sensor, composed of counters and an embedded antifuse memory block, is also presented to identify recycled ICs. The counters are used to record the usage time of ICs, and the value is then dynamically stored in the antifuse memory block by controlling the programming signal. Since the antifuse memory block is one-time programmable, "recyclers" cannot erase the context during the recycling process. Therefore, the AF-based sensor is resilient to removal and tampering attacks. Two different structures of AF-based sensor are proposed to measure the usage time of ICs in [19]: (i) an AF-based sensor using clock (CAF-based) records the cycle count of the system clock during the chip operation. The usage time of recycled ICs can be reported by this sensor and the measurement scale and total measurement time can be adjusted according to the application of ICs. (ii) an AF-based sensor using signal transition (SAF-based) selects a certain number of signals with low switching probability and records their switching activities to calculate usage time to detect recycled ICs with less area overhead compared to the CAF-based sensor.

10.1 Background

In this section, aging phenomenon in ICs and their impact on different circuit components will be briefly introduced. The antifuse OTP memory used in the AF-based sensor will also be briefly introduced.

10.1.1 Aging Analysis

When the chip operates in functional mode, the transistors age mainly due to NBTI and HCI. The aging effects of NBTI and HCI can cause parametric shifts and circuit failures, as demonstrated by reliability models [20,21], and [22]. NBTI occurs when a negative gate-to-source voltage is applied at the PMOS transistors, which breaks Si-H bonds generating the interface traps. These interface traps can increase the absolute value of the PMOS threshold voltage (V_{th}), resulting in reduced transistor current and increased gate delay. Equation (10.1) shows the shift of V_{th} caused by NBTI [23].

$$\Delta V_{th} = \frac{qN_{it,NBTI}(t)}{C_{ox}} \tag{10.1}$$

where C_{ox} represents the gate oxide capacitance, q is the electronic charge, and $N_{it,NBTI}(t)$ is the number of interface traps, which will increase as the transistors continue to operate in the field. HCI occurs when the electron or hole in the transistors gains sufficient energy to overcome the silicon dioxide barrier in order to break an interface state. The silicon substrate/gate dielectric interface and dielectric bulk traps caused by HCI can impact device parameters including threshold voltage, as shown in (10.2).

$$\Delta V_{th} = \frac{qN_{it,HCI}(t)}{C_{ox}} \tag{10.2}$$

where $N_{it,HCI}(t)$ is the number of interface traps caused by HCI.

Since recycled ICs have been impacted by these aging effects when used in the field, the circuit parameters of recycled ICs would be different from those of new ICs. If a *fast-aging sensor* was embedded into the circuit to help detect its usage, then recycled ICs could be identified.

In order to verify the effects of aging on a circuit's performance, several different inverter chains were simulated using Synopsys 90 nm technology [24]. The delay

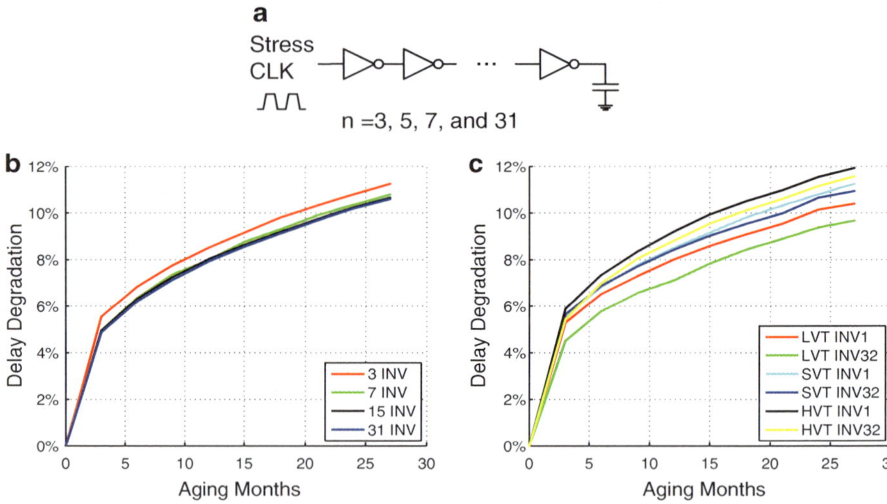

Fig. 10.3 (**a**)An inverter chain structure, (**b**) The degradation of inverter chains with different lengths (stage count), and (**c**) The degradation of a 3-inverter chain with different inverter types

of these inverter chains will represent the circuit's performance. The simulation was conducted using HSPICE MOSRA (Synopsys' reliability analysis tool) with combined NBTI and HCI aging effects at 25°C. Figure 10.3a shows the basic structure of the inverter chains with the same capacitive load and the same stress coming from a 500 MHz clock. These chains are composed of 3, 7, 15, and 31 standard, high, and low threshold voltage (SVT, HVT, and LVT) inverters. Figure 10.3b presents the delay degradation of inverter chains under clock stress for up to 27 months with no interruption. As can be seen in the figure, the number of inverters does not have a significant impact on the degradation of these chains since they receive the same stress, and each inverter's speed degrades at the same rate. Aging effects are also dependent on a device's threshold voltage. The 3-inverter chains were simulated using SVT, HVT, and LVT and two different size inverters (INVX1 and INVX32). Figure 10.3c shows that the chain with the HVT inverters experiences more degradation than the chains with the SVT or LVT inverters, and the INVX1 inverter chain has a larger degradation than the INVX32 inverter chain.

NAND and buffer (BUF) gate chains with HVT were also simulated at 25 °C with a 500 MHz clock stress. The basic structure of these chains is the same as the inverter chains. A NAND gate will function as an inverter when its two inputs are connected together. Figure 10.4 shows the simulation results. As can be seen in the figure, the gate type does not impact the aging speed significantly. However, the inverter chain ages slightly faster than the others, while the NAND gate chain and the BUF chain age at almost the same speed. The difference in the amount of aging depends on the structure of gates. Therefore, inverters (INVX1) with HVT will be used to create the ring oscillators used to detect recycled ICs in this simulation.

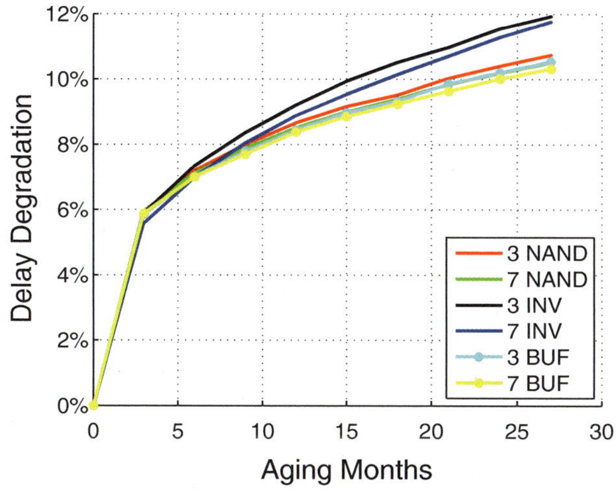

Fig. 10.4 The delay degradation of NAND, BUF, and INV chains

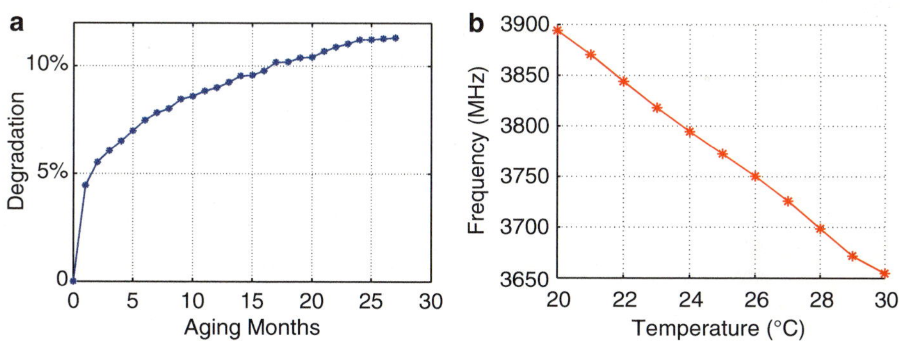

Fig. 10.5 (**a**) The frequency degradation of a 5-stage RO, and (**b**) The frequency of a 5-stage RO decreases as temperature increases

Figure 10.5a shows the frequency degradation of a 5-stage ring oscillator with HVT inverters after 27 months of aging. The frequency of the RO in a recycled IC will be smaller than in a new IC. If there are no environmental or process variations, recycled ICs could be easily identified by measuring the frequency of the RO embedded in the circuit. However, variations have a significant impact on the ROs' frequency. Figure 10.5b shows that the frequency of the 5-stage RO will decrease as the temperature increases, and that the frequency variation could be very large. Note that increasing temperature can also increase circuit degradation.

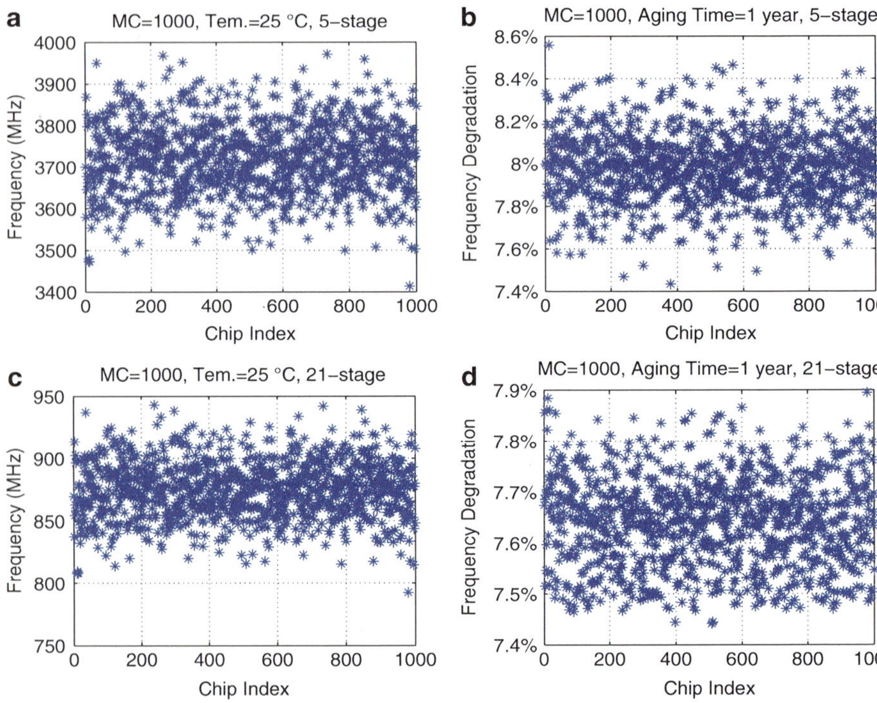

Fig. 10.6 (**a**) The frequency of a 5-stage RO varying with process variations, (**b**) The frequency degradation of a 5-stage RO aging for 1 year varying with process variations, (**c**) The frequency of a 21-stage RO varying with process variations, and (**d**) The frequency degradation of a 21-stage RO varying with process variations

The 1,000 Monte Carlo (MC) simulation results of the 5-stage RO are shown in Fig. 10.6a, at a temperature of 25°C with 3σ: 2 % Tox, 5 % Vth, and 5 % L inter-die variations and 1 % Tox, 5 % Vth, and 5 % L intra-die variations. As this figure displays, the RO's frequency can vary as much as 20 % under process variations. In addition, process variations impact the aging rate of the RO, as shown in Fig. 10.6b. The frequency degradation of the 1,000 chips varies around 8 % (7.4 %–8.6 %) for 1 year of aging. This frequency shift caused by the aging effects in recycled ICs can help separate them from those caused by process variations in new ICs.

With a fixed stress, the number of inverters does not have a significant impact on an inverter chains' delay degradation. However, the frequency of an RO is related to the number of inverters, $f = \frac{1}{2 \times n \times t_d}$, where n is number of stages in the RO and t_d is the delay of an inverter. Figure 10.6c shows the frequency shift of a 21-stage RO with HVT inverters. The frequency degradation is shown in Fig. 10.6d. It can be concluded that the 5-stage RO experiences slightly more degradation since its oscillation frequency is higher than the 21-stage RO by comparing the frequency degradation of the 5-stage and 21-stage ROs. However, a 5-stage RO may require a very fast counter that might be difficult to design for timing closure.

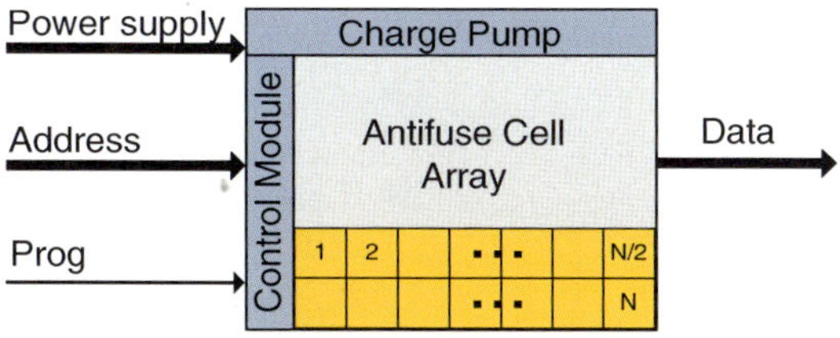

Fig. 10.7 The typical interface of antifuse memory

10.1.2 *Antifuse Memory*

An antifuse is an electronic device that changes the state from non-conducting/high resistance to low resistance in response to electrical stress. With sufficiently high voltage or current, a large power dissipation in a small area will melt a thin insulating dielectric between polysilicon and diffusion electrodes and form a thin, permanent, and resistive silicon link. The programming performed after manufacturing is irreversible and permanent in antifuse cells, which will be used in the AF-based sensor to store the usage time of ICs.

The AF-based sensor is composed of counters with the usage time of ICs when powered on stored in an embedded antifuse OTP memory block during the chip operation. Otherwise, the data may be erased or altered by attackers while the chip is in the power-off mode. The reasons for using an antifuse block in the AF-based sensor are [25]: (i) it consumes less power to program or read compared with other types of OTP structures, such as electrical fuse or CMOS floating gate, (ii) the area of an antifuse is much smaller than an efuse, and (iii) it does not require additional mask or manufacturing handling steps during fabrication.

However, most antifuse memories are programmed in a programming environment with relatively high voltage/current. Therefore, integrated charge pumps or voltage multipliers are used to provide sufficiently high voltage/current [26, 27] in embedded antifuse OTP memories. With those charge pumps or voltage multipliers, no additional power supply is required during programming. The typical interface of the embedded antifuse memory is shown in Fig. 10.7 [26, 27], including *Power supply, Address, Prog*, and *Data* signals. Existing antifuse blocks with the interface shown in Fig. 10.7 are used instead of designing a new embedded antifuse structure in the AF-based sensor, since embedded antifuse memory is only a small part of the sensor.

10.2 Recycled-IC Detection Sensors

Two different sensors to identify recycled ICs are presented in this chapter. The RO-based sensor is based on the aging differences between two ring oscillators to record the usage time of ICs. It does not require any memory element to store the usage time since it is hidden in the degraded RO frequency because of aging. AF-based

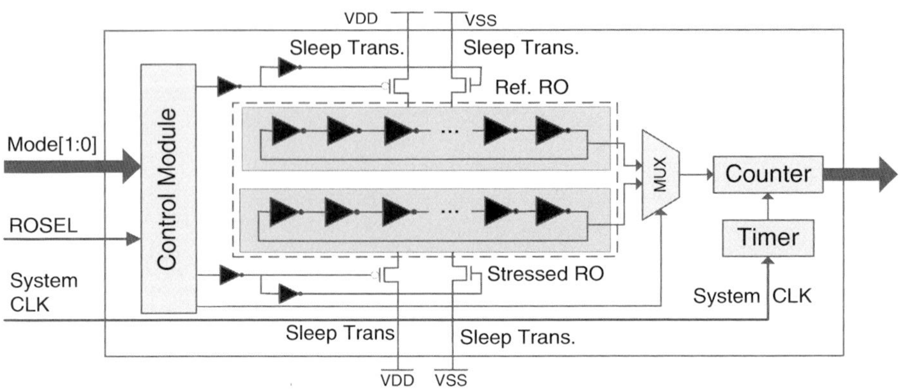

Fig. 10.8 The structure of the RO-based sensor

sensors count the system clock or the switching activity of the signals in the design and store the usage time in an antifuse OTP block. The two sensors will be discussed in detail in the following.

10.2.1 RO-Based Sensor

The main objectives in designing the RO-based sensor are: (i) the sensor must age at a very high rate to help detect ICs used for a short period of time, (ii) the sensor must experience no aging during manufacturing testing, (iii) the impact of process variations and temperature on the RO-based sensor must be minimal, (iv) the sensor must be resilient to attacks, and finally, (v) the measurement process must be done using low-cost equipment, and be very fast and easy.

As mentioned earlier, aging effects can slow down the frequencies of ROs embedded into the ICs. With an embedded RO, these recycled ICs can be identified based on their frequency, which will be lower than that of a new IC. However, there are many parameters impacting the frequency of an RO, such as temperature and process variations. The RO-based sensor uses a Reference RO and a Stressed RO to separate the aging effects from process/environmental variations.

Figure 10.8 shows the structure of the RO-based sensor, which is composed of a control module, a Reference RO, a Stressed RO, a MUX, a timer, and a counter. The counter measures the cycle count of the two ROs during a pre-specified time period, which is controlled by the timer. System clock is used in the timer to minimize the measurement period variations due to circuit aging. The MUX selects which RO is going to be measured, and is controlled by the $ROSEL$ signal. The Reference and Stressed ROs are identical; both are composed of HVT components. The inverters in Fig. 10.8 can be replaced by any other type of gates (NAND, NOR, etc) that can construct an RO. It will not change the effectiveness of the RO-based sensor significantly. Smaller-stage ROs are used in the RO-based sensor considering the counter's measurement speed limits given a technology. For example, in the 90nm technology, a 16-bit counter can operate under a frequency of up to 1GHz; an inverter-based RO of at least 21 stages is then required.

Sleep transistors are used to connect the ROs to the power supply in the RO-based sensor; PMOS sleep transistors control the connection between VDD and the inverters and the NMOS sleep transistors control the connection between VSS and the inverters. Both the Reference RO and the Stressed RO work in three modes controlled by the $Mode$ signal: (i) when the IC is in manufacturing test mode, the Reference RO and Stressed RO will be disconnected from the power supply and experience no aging. This mode only lasts a short time, depending on the IC test procedures. (ii) when the IC is in normal functional mode, the Reference RO will be disconnected from VDD and VSS, but the Stressed RO will be gated on and will age. The frequency of the Stressed RO will drop while the Reference RO will not change very much. ICs will spend most of their time in this mode. (iii) when the IC is in authentication mode (i.e., when an IC is taken from the market, and its authenticity is to be verified), both the Reference RO and Stressed RO will be gated on by connecting to the power supply. The timer and counter will be enabled to measure ROs' cycle count and the $ROSEL$ signal will select which RO to measure. The rest of the IC functionality would be turned off by the $Model$ signals and the authentication process takes a very short period of time. The three modes of operation ensure that (i) the frequency difference between the Reference RO and Stressed RO will be larger over time since the Reference RO cannot be gated on alone, and (ii) it is extremely difficult for adversaries to force the RO-based sensor to operate in authentication mode when it is supposed to be in its normal functional mode, which would eliminate the aging difference. The only method to do that would be to modify the original RO-based sensor module, which is impossible during a simple recycling process.

As shown in Fig. 10.8, the inverters of the Reference RO and the Stressed RO are placed next to each other physically, designed as a single small module. The process and environmental variations between them should be very small. Therefore, in a new IC, the frequency difference between the Reference RO and the Stressed RO would be within a certain small range. In a recycled IC, the Stressed RO will have suffered aging from its own oscillation, since the chip has been working in normal functional mode for a long time. However, the Reference RO will not have experienced as much aging since it was gated off. The frequency difference between the Reference RO and the Stressed RO will grow larger as the chip operates longer, which is demonstrated by the simulation and silicon results. If the frequency difference is outside of the new ICs' frequency difference range considering process variations, it can be concluded with high confidence that the CUT was recycled from used boards. The area overhead of the RO-based sensor is negligible when compared to the millions of gates in modern ICs. Power consumption is also limited to that consumed by the Stressed RO in the RO-based sensor.

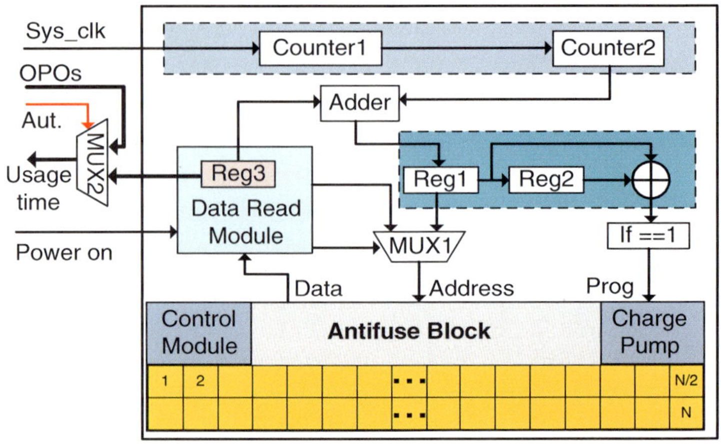

Fig. 10.9 The structure of the CAF-based sensor

10.2.2 AF-Based Sensor

In the RO-based sensor, the inverters of the Reference RO and the Stressed RO are placed physically next to each other to minimize the impact of intra-die process variations. However, it may still be difficult to completely exclude the impact of inter-die process variations on the sensor. In addition, the RO-based sensor provides only an approximation of the usage time in a form of aging in the stressed RO. Therefore, the sensitivity (the minimum usage time of recycled ICs detected by sensors) of the RO-based sensor is limited. For example, it may not identify recycled ICs used for less than 1 month, based on simulation. In order to eliminate the issue of process variations, to provide a more accurate usage time, and to identify recycled ICs that are only used for a very short period of time (such as 1 day), two AF-based sensors are proposed: the CAF-based sensor and the SAF-based sensor.

CAF-Based Sensor: Fig. 10.9 shows the structure of the CAF-based sensor, which is composed of two counters, a data read module, an adder, and an antifuse OTP memory block. *counter1* is used to divide the high frequency system clock with a lower frequency signal, as shown in Fig. 10.9. *counter2* is used to measure the cycle count of the lower frequency signal. The size of the two counters can be adjusted depending on the measurement scale (T_s: defined as the time unit reported by the sensor) and the total measurement time (T_{total}). For example, if T_s is 1 h and T_{total} is 1 year based on the specification of an IC, a 38-bit *counter1* will meet the requirement to count the usage time from 20ns (assume system clock = 50 Mhz) to 1h and a 14-bit *counter2* will count the usage from 1 to 8,760h (1 year).

Since the data stored in registers (counters) could be lost or reset when the power supply is off, non-erasable memory is required in this sensor. An embedded antifuse OTP block is used instead of a field programmable read-only memory (FPROM)

Fig. 10.10 The algorithm for "data read" in CAF-based and SAF-based sensors

Algorithm for Data Read

```
initial address = (N/2);
for (i = log(N/2),i > 0,i−−) {
   if ([address] == 1)
      address = address + 1;
      if ([address] == 0)
         address = address −1, $stop;
      else
         address = address − 1;
      address = address + 2^(i −1);
   else
      address = address − 2^(i +1);
}
```

to store the usage time information because FPROM could be tampered with or altered by attackers. In the antifuse block, *Prog* is assigned to be $1'b1$ if the value in *counter2* increases by "1". By connecting the output of *counter2* to *Address* in the antifuse block directly, the related antifuse cell will be programmed as "1". Therefore, the largest address of the cell whose content is "1" will be the usage time of CUT based on the measurement scale setup by *counter1*.

However, program and read operations share the same *Address* signals in an antifuse block. Therefore, a MUX (*MUX1* in Fig. 10.9), controlled by *data read module*, is used to select the address (antifuse cell) to be read or programmed. Every time the power supply is on, the antifuse block will work in read mode for a short period of time. During this time, the read address generated by the *data read module* will go through *MUX1*, and all the antifuse cells will be traversed based on the traversing binary tree principle. Figure 10.10 shows the algorithm for data read in an N-bit antifuse block. As Fig. 10.10 illustrates, there are $log(N/2)$ loops in the algorithm. The address is increased or decreased by $2^{i-1}(i = 0, \ldots log(N/2))$ for the ith loop based on the value in the address. If the value stored in the address is "1", ($[address] == 1$) and the value stored in the next address is "0", the address will represent the usage time before power-on, based on T_s. The read operation will last less than $log(N/2) + 1$ system clock cycles, depending on the value stored in the antifuse block; this time will be recorded by *counter1*, as well.

Once the previous usage time is collected, it will be stored in register *Reg3* and sent to the $adder$. The reason for using an adder here is that counters start from "0" every time the power is turned on, and the previous usage time must be considered when the total usage time is calculated. In addition, *Reg1* is used to sample the data in *adder*, *Reg2* delays the data in *Reg1* with one system clock, and *XOR* gates are used to compare the data in *Reg1* and *Reg2*. If they are different (denoting the usage time increased), the antifuse OTP block will work in program mode and the data in *Reg1* will go through *MUX1* to the $Address$ in the antifuse block. Therefore, combined with the value in *counter2* (the usage time after power-on), the new total usage time will be stored in the antifuse OTP block by programming a new antifuse

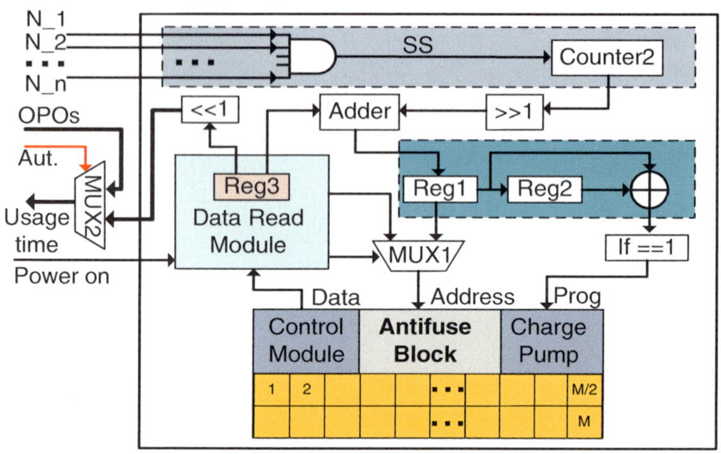

Fig. 10.11 The structure of the SAF-based sensor

cell with a larger address. By designing the sensor in this way, the probability of altering or tampering attacks will be reduced on the AF-based sensor.

In order to eliminate the need for additional pins on the chip for authentication purposes, the CAF-based sensor uses a MUX (*MUX2*) and an authentication (*Aut.*) pin to send the usage time to the output pins of the ICs. This way, no extra output pins will be added to the original design. When the IC works in normal functional mode, original primary outputs (*OPOs*) will go through *MUX2*. If the IC is in authentication mode by enabling the authentication signal, the *data read module* will set the antifuse IP in read mode and the usage time will go through *MUX2*. In addition, when the IC works in the manufacturing test mode, the functionality of the CAF-based sensor will be disabled, and structural fault test patterns will be applied to the sensor.

SAF-Based Sensor: With two counters, the area overhead of the CAF-based sensor could still be considered large for smaller designs. In order to reduce the area overhead, an SAF-based sensor is proposed based on the (SW), as shown in Fig. 10.11. Comparing Figs. 10.11 with 10.9, it can be seen that the structure of the SAF-based sensor is similar to that of the CAF-based sensor. The difference is that the CAF-based sensor counts the cycle of system clock to record the ICs' usage time while the SAF-based sensor counts the switching activity (positive edge) of a certain number of nets in the design. During simulation, a certain number of nets are selected to be the input of an AND gate. The rule of net selection is that the switching activity of the output of the AND gate must meet the requirement of the measurement scale. For example, if T_s is 1h, one of the choices could be four nets with $SW(N\_1) = 30/60$ min, $SW(N\_2) = 24/60$ min, $SW(N\_3) = 25/60$ min, and $SW(N\_4) = 24/60$ min, respectively. However, with different functional inputs, the signals' SW could be significantly different. Therefore, only the signals with consistent SW under different inputs are selected. The net selection can be adjusted based on different designs and measurement scales. Then the positive pulse of the output of the AND gate (SS signal in Fig. 10.11) will be counted by *counter2* in the sensor.

In order to further reduce the area overhead, a 1-bit right shifter is used to divide the value in *counter2* by 2, and then the largest address of antifuse cells with “1” will represent *[SW/2]*. A 1-bit left shifter is used to calculate the switching activity by $[SW/2] \times 2$. The recorded *SW* will represent the ICs' usage time. Therefore, the number of antifuse cells in SAF-based sensor will be reduced compared with CAF-based sensor. However, the accuracy of the SAF-based sensor is lower than the CAF-based sensor because (i) it is based on the switching activity of a certain number of nets on the netlist while CAF-based sensor counts the cycle count of the system clock, and (ii) the SAF-based sensor loses part of the usage time information due to the shifters.

Compared with the RO-based sensor, the area overhead of the two AF-based sensors is larger because of the counters and the antifuse OTP block. However, this overhead is still negligible when compared to the millions of gates in modern ICs. The major advantage of the AF-based sensor over the RO-based sensor is that the usage time stored in the AF-based sensors to identify recycled ICs will not be impacted by technologies (i.e., older technology designs do not age as much as the new ones do), packages, assemblies, or process variations. Even if the design was fabricated at different times in different foundries, the AF-based sensor can still indicate how long a chip under testing has been used. In addition, AF-based sensors can identify recycled ICs used for a very short period of time, such as 1 day, due to the small measurement scale.

10.3 Results and Analysis

In this section, the experimental results of the RO-based sensor and AF-based sensor will be presented, including simulation results and silicon results from test chips. Attack analysis on the two sensors will also be discussed.

10.3.1 RO-Based Sensor

Figure 10.12 shows the proposed measurement flow using the RO-based sensor for identifying recycled ICs in experiments. This is done only for the purpose of validating the proposed sensor. The way the RO-based sensor is designed eliminates the need for a golden IC, especially when a chip is used in the field for a long period of time in the field. First, a certain number of random, new ICs are used as sample chips to generate a fingerprint. The samples can come from the same or different wafers and lots. The larger this sample is, the more

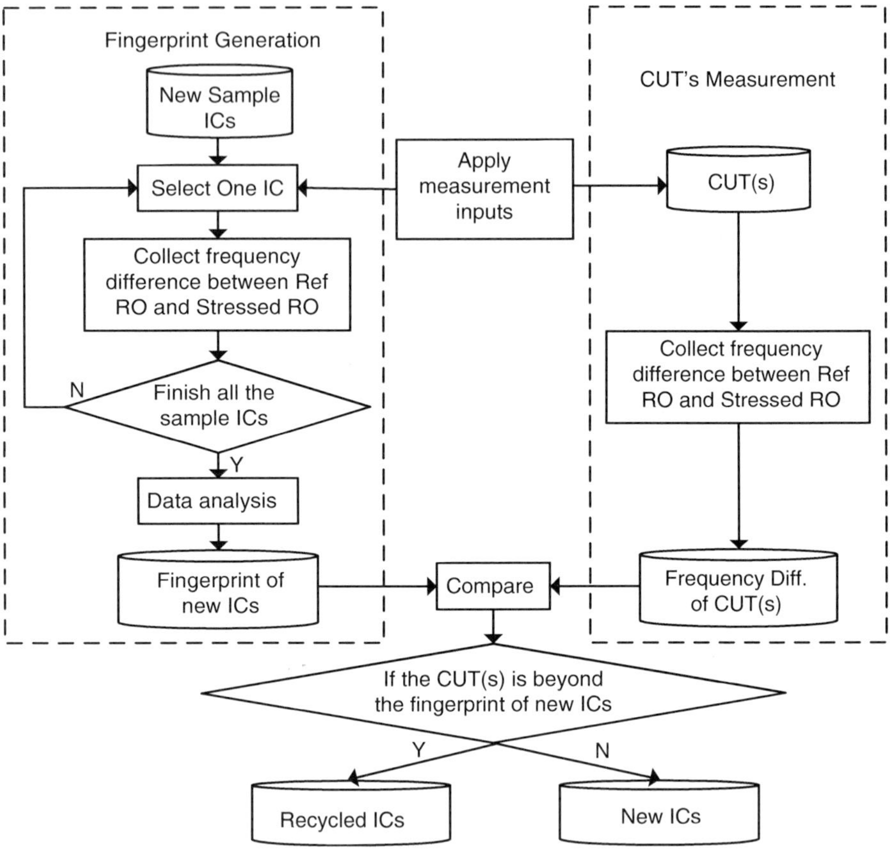

Fig. 10.12 The measurement flow using the RO-based sensor for identifying recycled ICs

process variation space will be covered, reducing the probability that new ICs with large process variations will be identified as recycled ICs. 1,000 sample chips are tested in the simulation. In authentication mode, the Reference RO's and Stressed RO's frequencies are measured. Temperature variation should not impact the identification results significantly, since the Reference RO and Stressed RO will experience the same environmental temperature.

Once the sample chips have been measured, the frequency difference between the Reference RO and Stressed RO will be calculated, with $F_{diff} = F_{ref} - F_{str}$, where F_{ref} is frequency of the Reference RO, and F_{str} is frequency of the Stressed RO. With 1,000 sample chips, the range of F_{diff} will be determined using distribution analysis, creating a fingerprint for new ICs. If F_{diff} of the CUT is out of the range of the new ICs' fingerprint, there is a high probability that the CUT is a recycled IC. Otherwise, the CUT is assumed to be a new IC. The longer the CUT has been used, the more aging effects it will have experienced, making it easier to identify.

Table 10.1 Process variations

	Inter-die			Intra-die		
	Vth (%)	L (%)	Tox (%)	Vth (%)	L (%)	Tox (%)
PV0	5	5	2	5	5	1
PV1	8	8	3	7	7	2
PV2	20	20	6	10	10	4

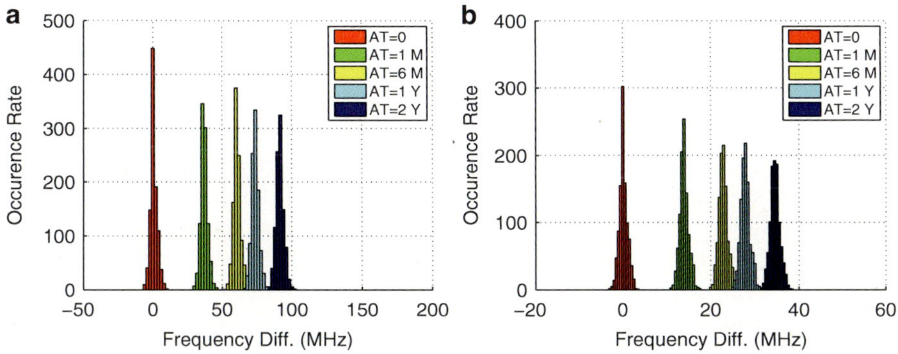

Fig. 10.13 The frequency difference distribution of RO-based sensor with PV0 using (**a**) 21-stage ROs, and (**b**) 51-stage ROs

The entire measurement procedure for each CUT should take only a short amount of time (less than few seconds).

90nm technology is implemented and simulated to verify the effectiveness of the RO-based sensor [24]. HSPICE MOSRA from Synopsys is used to simulate and measure the impact of aging on the RO-based sensor. The nominal supply voltage is 1.2V. During simulation, in the stress phase, the Reference RO was gated off and the Stressed RO was gated on, experiencing NBTI and HCI aging. The stress for the Stressed RO comes from its own oscillation. In the authentication phase, the Reference RO and Stressed RO were both gated on and measured one by one, selected by the $ROSEL$ signal. The measurement time was set up in the timer as 100μs in the simulation. Since the clock of the counter in the RO-based sensor is from the RO, the counter gives the cycle count of each RO. The frequency of RO is equal to the cycle count divided by the measurement time. The following simulation analysis is based on inverter ring oscillators.

Stage Analysis: RO-based sensors with 21-stage and 51-stage ROs were simulated at 25°C with 2 % Tox, 5 % Vth, and 5 % L inter-die and 1 % Tox, 5 % Vth, and 5 % L intra-die process variations (PV0 in Table 10.1). 1,000 chips were generated using Monte Carlo simulation by HSPICE, and the total aging time was set at 24 months with a 1-month step.

Figure 10.13a shows the frequency difference F_{diff} range between the 21-stage Reference RO and Stressed RO, where, in the legend, AT denotes aging time, M

represents month, and Y represents years. As seen in this figure, the frequency difference in new ICs ($AT = 0$) could be larger or smaller than 0, which is dependent on the process variations between the two ROs. In addition, the process variations of the CUTs are different from that of the 1,000 sample new ICs, but the frequency differences still follow an identical distribution. The range of frequency differences in the new sample ICs is used as the fingerprint. After being used for 1 month, the Stressed RO suffered from aging effects and its frequency became lower. The lowest frequency difference between the Reference RO and the Stressed RO is larger than the largest frequency difference present in the new IC set. Therefore, the recycled IC detection rate for ICs aged for 1 month or longer is 100 %. At 6 months, 1 year, and 2 years, the frequency difference between the Reference RO and the Stressed RO becomes larger and larger, and the variation of the frequency difference becomes larger as well. This is because the aging rate is different from chip to chip due to process variations; some ICs aged faster and some others aged slower.

RO-based sensors with 51-stage ROs were also implemented using the same temperature and the same process variations. Figure 10.13b shows the simulation results. By comparing Fig. 10.13a and b, it can be concluded that the frequency difference between aged and new ICs is smaller when using larger-stage ROs. However, the frequency difference variation becomes smaller as well, which means that the RO-based sensor could still detect fully recycled ICs that had been used for 1 month with a 100 % detection rate. If the RO-based sensor uses large-stage ROs, it may impact the absolute value of the frequency difference between the Reference RO and the Stressed RO, but the detection rate will not be impacted significantly. For different technologies, the stage count of the ROs could be adjusted based on the speed of the counter. In the following, RO-based sensors with 21-stage ROs are used for further analysis according to the 90nm technology.

Process Variations and Temperature Analysis: The effectiveness of the RO-based sensor is partly dependent on the variations between the Reference RO and the Stressed RO. With lower rates of variation, the RO-based sensor can identify recycled ICs that have aged for a shorter period of time. However, the variations between the Reference RO and the Stressed RO are determined by intra-die process variations. The smaller the intra-die variations, the more effective the RO-based sensor will be. Table 10.1 shows the different process variation rates used in the simulation to analyze their impact on detection. Moving from PV0 to PV2, inter-die and intra-die variations both become larger. RO-based sensors with 21-stage ROs were simulated at 25°C using these process variation rates.

By designing the sensor as a small module (hard macro), the Reference RO and the Stressed RO were placed close to one another and the variations between them were minimal. The simulation results of 1,000 chips with PV1 and PV2 are shown in Fig. 10.14a and b, respectively. By comparing Figs. 10.13a, 10.14a, and b, it can be concluded that the variation of the frequency differences between the Reference RO and the Stressed RO in the new ICs becomes larger with larger process variations. For the 1,000 ICs with PV2, the detection rate of recycled ICs aged for 1 month is 95.2 %. However, for the recycled ICs that aged for 6 months, the detection rate

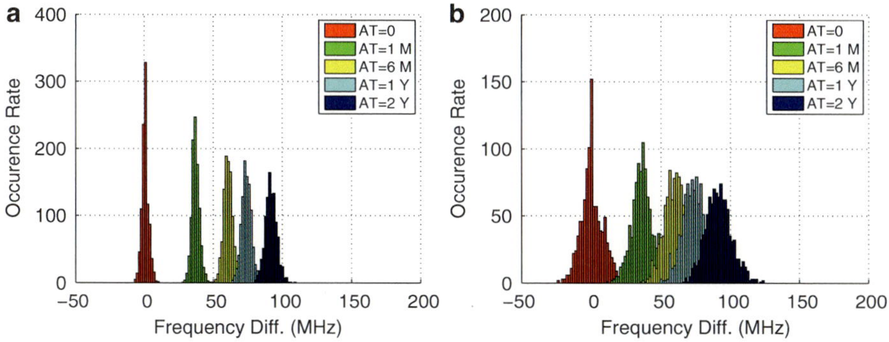

Fig. 10.14 The frequency difference distribution of an RO-based sensor with 21-stage ROs with (**a**) PV1 and (**b**) PV2

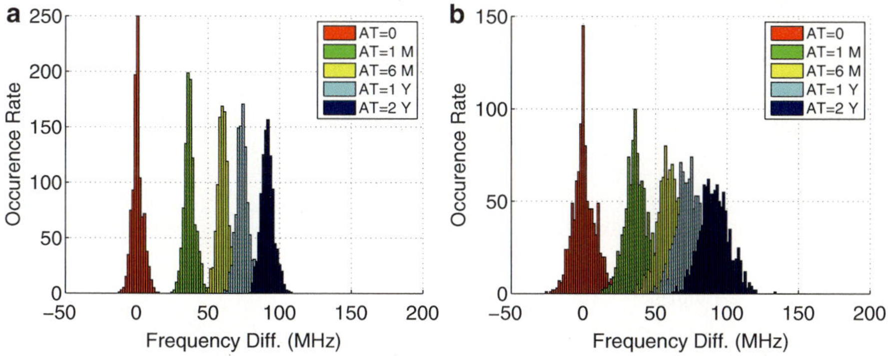

Fig. 10.15 The frequency difference distribution of the RO-based sensor with (**a**) PV1 and ±10°C and (**b**) PV2 and ±20°C

is 100 %. The RO-based sensor identifies shorter-aged recycled ICs with smaller intra-die process variations as in PV0, PV1, and PV2.

The 1,000 circuits generated using Monte Carlo were also simulated with both process and temperature variations. Figure 10.15a shows the frequency difference occurrence rate between the 21-stage Reference and Stressed ROs with process variations PV1 (shown in Table 10.1) and temperature variations of ±10°C around room temperature. Figure 10.15b shows the simulation results with process variations PV2 and temperature variations of ±20°C around room temperature. The results in Figs. 10.15a and 10.14a are from chips with the same process variations but different temperature variations. As the figures show, the frequency difference variations in Fig. 10.15a are slightly larger than those in Fig. 10.14a, due to temperature variations. The same conclusion can be made by comparing Figs. 10.15b and 10.14b. For the 1,000 chips with PV2 and ±20°C temperature variations, the detection rate of recycled ICs aged for 1 month is 92.3 %, but it is still 100 % for recycled ICs aged for 6 months, demonstrating that the RO-based sensor is effective even

Table 10.2 The structure of RO-based sensors in the test chip

	ROs in RO-based sensors			
	Reference RO	Stressed RO	RO Structure	Vth
RO-based1	R_RO1	S_RO1	1 NAND + 200 BUFs	SVT
RO-based2	R_RO2	S_RO2	1 NAND + 200 BUFs	HVT
RO-based3	R_RO3	S_RO3	201 NANDs	HVT

with large process and temperature variations. Note that such a large variation in temperature and process are not expected in practice when authenticating a CUT. The temperature difference and process variations between the two ROs in an RO-based sensor will be negligible since they are placed physically near each other.

Silicon Results: The RO-based sensor is also verified through analysis of test chips fabricated using a 90nm technology. The test chip was originally designed to verify the effects of aging on the frequency of the ROs. It is used here to demonstrate the effectiveness of the RO-based sensor. In total, there are 96 delay chains in the chip that can work in ring oscillator mode by controlling different input signals [28]. Six of these ring oscillators were selected to construct three RO-based sensors, as shown in Table 10.2.

- RO-based1 contains two identical ROs (R_RO1 and S_RO1) with one SVT NAND gate and 200 SVT BUFs;
- RO-based2 is composed of two identical ROs (R_RO2 and S_RO2) with one HVT NAND gate and 200 HVT BUFs
- RO-based3 includes ROs (R_RO3 and S_RO3) with 201 HVT NAND gates.

where R_RO1, R_RO2, and R_RO3 are Reference ROs, while S_RO1, S_RO2, and S_RO3 are Stressed ROs, respectively.

Comparing ROs included in the test chip with those used for HSPICE simulation, there are two main differences: (1) the stage of ROs in the test chip is 201 while the stage of ROs used in Monte Carlo simulation is much smaller (e.g. 21). The much larger number of stages in the test chip was used to make the measurement and observation possible with low-end oscilloscopes. (2) the gates in the ROs in the test chip are complex gates (BUFs, NANDs, etc.) while inverter-based ROs were used in simulation. The purpose is to aim at analyzing the impact of aging on different types of gates in the test chip. However, the number of stages and RO gate type do not present a significant impact on the effectiveness of the RO-based sensor.

Fifteen test chips are used in the experiment to present the impact of process variations and aging. To replicate the RO-based sensor's stressed mode, S_RO1, S_RO2, and S_RO3 were enabled and experienced accelerated aging for 80 h at 135°C with an elevated supply voltage (1.8V instead of 1.2V). The reason for using accelerated aging is that it takes a long time (usually weeks/months) to observe aging effects under normal conditions. The remaining three ROs were gated off and experienced no aging. In authentication mode, all of the ROs were enabled, and the temperature was brought back to room temperature (around 25°C). With the 15

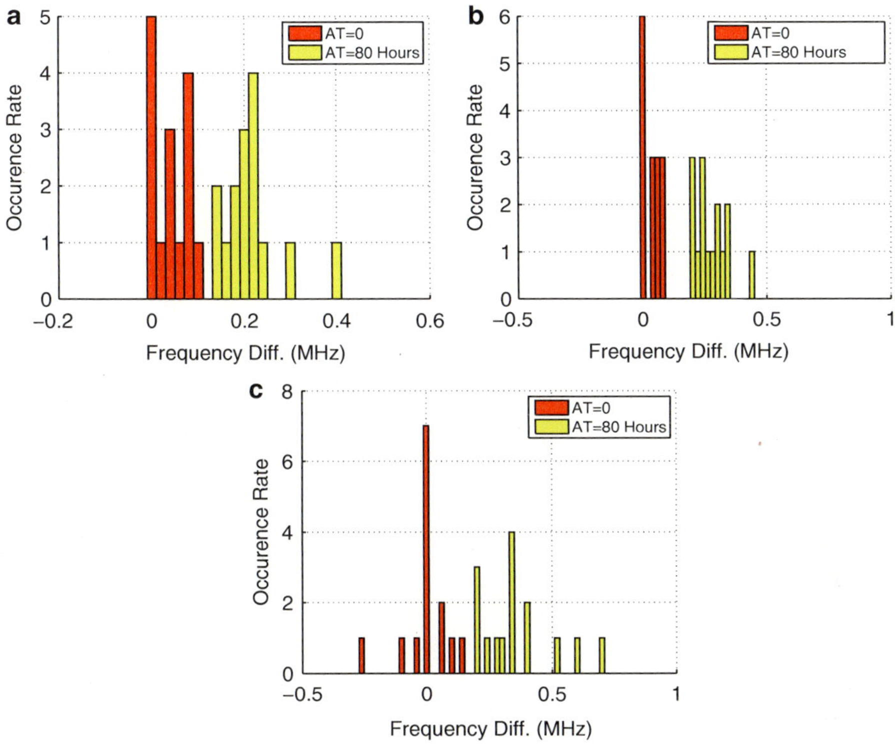

Fig. 10.16 The frequency difference distribution in (**a**) RO-based1, (**b**) RO-based2, and (**c**) RO-based3

new test chips, the average frequency of ROs is about 7.5 Mhz. Figure 10.16 shows the experimental results of the three RO-based sensors over the test chips. The red bars in the figure show the frequency difference between the Reference RO and the Stressed RO in each RO-based sensor at time zero (new/unused ICs). Similarly, the yellow bars are the frequency difference between the two ROs after 80h of aging.

Since a much larger number of stages are used in these sensors compared to those used in the simulations, the mean frequency of the ROs in the test chip and the frequency difference values are quite different from those yielded by simulations. However, even with 201 gates in these ROs, the detection rates of the recycled ICs that aged 80h using RO-based1, RO-based2, and RO-based3 are all still 100 %, which demonstrates that the RO stage count in the RO-based sensor does not have a significant impact on the sensor's effectiveness in detecting recycled ICs. According to the detailed results, the average frequency degradation of the stressed ROs in RO-based1, RO-based2, and RO-based3 (shown in Fig. 10.16) is 3.2 %, 4.0 %, and 3.8 %, respectively, By comparing Fig. 10.16a and b, it can be concluded that the frequency difference gap between new chips and aged chips in RO-based2 is larger than that in RO-based1. This is due to the fact that RO-based sensors with HVT gates

(RO-based2) will be more effective than those with SVT gates (RO-based1), which is also demonstrated in Fig. 10.16 through the simulation results. By comparing detection rates in Fig. 10.16b using RO-based2 (composed of HVT buffers) with Fig. 10.16c using RO-based3 (composed of HVT NAND gates), it can be concluded that the gates used in the RO can change the effectiveness of RO-based sensor slightly but not significantly.

Note that the ROs in the RO-based sensors in the test chip were not placed as close as they were supposed to. For instance, the results at time zero show that for RO-based1 and RO-based2, the R_ROs are faster than S_ROs in most cases, while this is not the case for RO-based3. This could be because of the spatial variations that exist between the ROs not placed near each other, which made some ROs faster than others. *For a RO-based sensor to be the most effective, it is recommended to place both ROs in a single localized module to reduce the variation between them.* Limited by the amount and structure of the test chips, the analysis with silicon data was not done as with the Monte Carlo simulations. However, the silicon results from these test chips demonstrate the effectiveness of the RO-based sensor.

10.3.2 AF-Based Sensors

From the above analysis, it can be concluded that detection of a recycled chip depends on the amount of degradation caused by aging, workload, and process and environmental variations. However, if the chip is used for a very short period of time, or if the chip is designed and fabricated using an older technology node, it will not experience much degradation, thus negatively impacting the effectiveness of detection. Process and temperature variations cannot impact the data in antifuse cells in the AF-based sensor since the usage time of the ICs is calculated by counters and stored in the antifuse block. Therefore, the only step required to know how long the IC has been used is to read the antifuse block by enabling the authentication signal. Note that a non-zero usage time from an AF-based sensor in a CUT does not suggest that it is a recycled IC due to the burn-in process. The CUT can be identified as a recycled one only if the usage time is longer than the time for the burn-in process. Therefore, recycled ICs used for a very short period of time can still be detected by the AF-based sensors.

Area Overhead Analysis: The area overhead of AF-based sensors are analyzed based upon the implementation of a design (named CSAFTEST) with about 500K gates and 12KB in-system programmable memory. Table 10.3 shows the area overhead caused by RO-based, CAF-based, and SAF-based sensors with different measurement scales and total measurement time. As can be seen in the table, the area overhead caused by the AF-based sensors changes with T_s and T_{total}, since the structure of AF-based sensors changes with measurement resolutions. For a CAF-based sensor, the size of *counter1* depends on T_s while the size of *counter2* and the size of the antifuse memory block both depend on T_{total}/T_s.

Table 10.3 The area overhead caused by RO-based, CAF-based, and SAF-based sensors on CSAFTEST

Measurement		Area overhead			
Scale (T_s)	Total Time (T_{total})	RO-based	CAF-based (%)	SAF-based (%)	Reduction (%)
1min	1 month	–	7.37	3.72	49.5
1h	1 year	–	1.57	0.82	47.8
1 day	1 year	–	0.18	0.12	33.3
1 day	4 years	–	0.37	0.21	43.2
–	–	0.025%	–	–	–

For SAF-based sensor, the area overhead is much smaller than that of CAF-based sensor due to the shifters. The reduction, calculated by *{Overhead(CAF-based)-Overhead(SAF-based)}/Overhead(CAF-based)*, is shown in the sixth column in Table 10.3. For example, with $T_s = 1$ h and $T_{total} = 1$ year (8,760h), CAF-based sensor was designed with 20-bit *counter1*, 14-bit *counter2*, and 8,760-bit antifuse memory block. The area overhead of this CAF-based sensor is 1.57 % while the area overhead caused by SAF-based sensor is 0.82 % and the reduction is 47.8 %. However, if $T_s = 1$ min &, $T_{total} = 1$ month, $T_s = 1$ day, and & $T_{total} = 1$ year, the area overhead of CAF-based sensor are 7.37 % and 0.18 %, respectively.

From the above analysis, it can be concluded that the area overhead caused by AF-based sensors depends on the application and specification of ICs. For example, if an IC is used in a system that requires a small T_s and a large T_{total}, the area overhead would be large. Otherwise, the overhead would be small (less than 1 %). On the other hand, the time recorded by the AF-based sensors is the power-on time and the intervals between power-on are not calculated. Therefore, the usage time stored in the sensor (T_{total}) is usually shorter than the time with power-off intervals. With a smaller T_{total}, the size of the antifuse memory block in the AF-based sensors will be smaller, and accordingly, the area overhead will be smaller.

Furthermore, the following conclusions could be made by comparing the RO-based sensor with AF-based sensors: (i) the area of an RO-based sensor is much smaller than that caused by AF-based sensor, and it also stays constant because the number of gates used in RO-based sensor does not vary with designs. Here, the RO-based sensor was about 0.025 % area overhead, which is negligible. (ii) the accuracy of RO-based sensor is lower than that of the AF-based sensors since it only provides an approximation of the usage time in a form of aging in the stressed RO.

Usage Time Analysis: Since the AF-based sensor only records usage time larger than T_s, if the power-on time of an IC is smaller than T_s, part of the usage time will be lost during the measurement. In order to verify the usage time, CAF-based and SAF-based sensors are analyzed with different T_s. Take the worst case for example, if every time the IC is turned on, the power-on time (T_{pon}) is shorter than T_s, then the AF-based sensors will not record any usage time. The value stored in the antifuse memory will always be equal to the time for the burn-in process. The AF-based sensors will be ineffective in this case, which should be avoided when designing an AF-based sensor.

With appropriate T_s, $N = [T_{pon}/T_s]$ will be recorded in *counter2* every time power is on and combined with the previous usage time to be stored in the antifuse memory block in CAF-based sensor. Figure 10.17a shows the estimated usage time under different usage situations using a CAF-based sensor. The X axis represents the worst case when $T_{pon} < T_s$. In this case, the estimated usage time recorded by the sensor is always zero. The solid line represents the ideal case, when the estimated usage time (T_{esm}) is equal to the actual usage time. The range between the dashed line and solid line represents the estimated usage time when $T_{pon} > T_s$. Similarly, the range between the dash-dot line and solid line represents the estimated time when $T_{pon} > 10 \times T_s$. As can be seen in the figure, the longer the chip is used on each power-on, the more accurate estimated usage time will be recorded by the CAF-based sensor.

For SAF-based sensor, the estimated usage time under different usage situations is shown in Fig. 10.17b. By comparing Fig. 10.17b with Fig. 10.17a, a conclusion can be made that the accuracy of the SAF-based sensor is slightly lower than that of CAF-based sensor. For example, when $T_{pon} > T_s$, the usage time recorded by CAF-based sensor would be $T_{est} = [T_{pon}/T_s] \times T_s$ while the usage time recorded by SAF-based sensor would be $T_{est} = [T_{pon}/2T_s] \times 2T_s$. In addition, since the SAF-based sensor is based on the switching probability of several nets in the netlist, the estimated usage time shown in Fig. 10.17b is based on a probability. Assuming that the output of the AND gate in the SAF-based sensor (SS signal in Fig. 10.11) switches once during T_s with probability p, then SS will switch more than once with probability $1 - p$. Note that the case that SS does not switch during T_s will not be considered since this situation should be avoided when designing an SAF-based sensor. With this assumption, when $T_{pon} > 2 \times T_s$, the estimated usage time will be in the range between the dashed line and solid line with probability p, shown in Fig. 10.17b.

Note that even with time lost during the measurement by using AF-based sensors, a recycled IC could still be identified since the usage time recorded by the antifuse memory block in the used ICs will be longer than the time for the burn-in process. After the burn-in process and before being sent to market, the AF-based sensor in all CUTs report almost identical usage time. However, when ICs are used in the field, the usage times recorded by the sensor in CUTs would be larger and different from each other based on the usage conditions before recycling.

10.3.3 Attack Analysis

In this section, a few attacks circumventing RO-based and AF-based sensors are discussed considering the capabilities of professional recyclers. The first attack to the RO-based sensor could be removal and tampering attacks. However, it is inherently difficult for the recycler to remove the sensor, due to the expected measurement results from the two ROs. The second attack could be that the recycler tries to intentionally age the Reference RO to mask the difference between the ROs

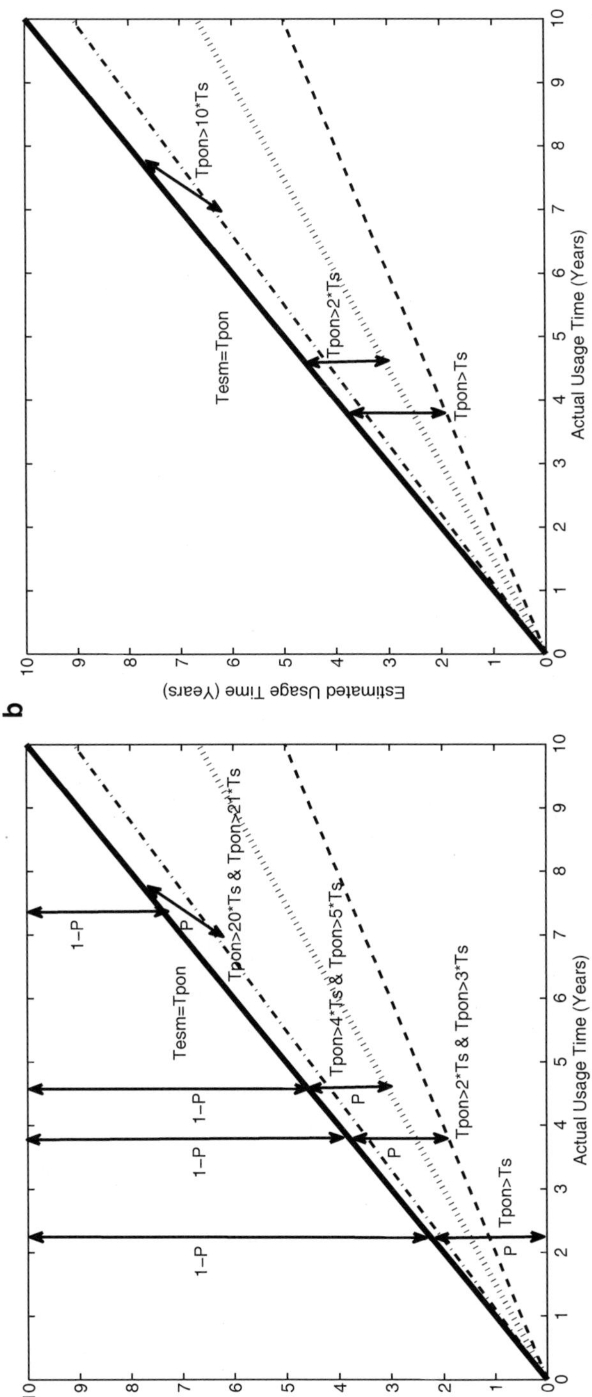

Fig. 10.17 Usage time analysis using (**a**) CAF-based sensor and (**b**) SAF-based sensor

in the RO-based sensor. Similarly, it is impossible to do that since Reference RO cannot be gated on alone. However, one can argue that attackers with unlimited resources may be able to remove the chip package, modify the original design, and tamper with the RO-based sensor. For such ICs where additional security is required, alterations can be made to the RO-based sensor to prevent these kinds of attacks. The RO-based sensor could be obfuscated inside the IC by multiplexing functional gates. This modification would make it more difficult for an attacker to analyze the IC, and make it more difficult to tamper with the sensor or modify it in any way.

For AF-based sensors, attackers would try to mask the usage time of ICs by disabling the sensor. However, the AF-based sensor will automatically run whenever the power is on, and the usage time will be stored in the antifuse memory directly. Therefore, it is impossible for attackers to disable the sensor without removing the packaging and breaking the chip. The second attack could be the erasure and alteration of antifuse cells; this is not possible because the memory used in the sensors is an antifuse OTP block. The most important advantage of the antifuse OTP technique is its ability to resist all existing reverse engineering methods because the oxide breakdown in antifuse cells occurs in a random location within a bounded enclosure and is extremely small [25]. Therefore, the state of a bit cell remains hidden in the silicon atoms, making it extremely difficult for attackers to tamper with the memory. The third attack could be modification of the counters or the signals connected in the sensor. However, with limited resources and without access to the original design, attackers cannot modify the nets' connection. Decapping, professional cleaning, and remarking will not help attackers either.

10.4 Summary

In this section, two techniques are presented using lightweight on-chip sensors to detect recycled ICs. The frequency difference between the Reference RO and the Stressed RO in the RO-based sensor makes the identification of recycled ICs possible. The usage time stored in the antifuse memory using AF-based sensors can indicate how long an IC has been used and then identify a recycled IC. Experimental results and analysis demonstrated the effectiveness of these sensors.

References

1. "Defense Industrial Base Assessment: Counterfeit Electronics," Bureau of Industry and Security, U.S. Department of Commence, *http://www.bis.doc.gov/defenseindustrialbaseprograms/osies/defmarketresearchrpts/final_counterfeit_electronics_report.pdf*
2. Business Week, "Dangerous Fakes," *http://www.businessweek.com/magazine/content/08_41/b4103034193886.htm,* 2008.
3. L. W. Kessler and T. Sharpe, "Faked Parts Detection," *http://www.circuitsassembly.com/cms/component/content/article/159/9937-smt,* 2010.

4. J. Stradley and D. Karraker, "The Electronic Part Supply Chain and Risks of Counterfeit Parts in Defense Applications," *IEEE Transactions on Components and Packaging Technologies*, pp. 703–705, Sept. 2006.
5. Military Times, "Officials: Fake Electronics Ticking Time Bombs," *http://www.militarytimes.com/news/2011/11/ap-fake-electronics-ticking-time-bomb-110811/*, 2011.
6. Tezzaron Semiconductor, "3D-ICs and Integrated Circuit Security," *http://www.tezzaron.com/about/papers/3D-ICs_and_Integrated_Circuit_Security.pdf*, 2008.
7. *http://www.combatcounterfeits.com/gallery.htm*.
8. "Counterfeit Electronic Parts; Avoidance, Detection, Mitigation, and Disposition", *http://standards.sae.org/as5553/*.
9. K. Lofstrom, W. R. Daasch, and D. Taylor, "IC Identification Circuit Using Device Mismatch," in *Proc. ISSCC*, pp. 370–371, 2000.
10. R. Pappu, "Physical One-way Functions," *Phd thesis, MIT*, 2001.
11. G. Suh and S. Devadas, "Physical Unclonable Functions for Device Authentication and Secret Key Generation," in *Proc. DAC*, pp. 9–14, 2007.
12. E. Ozturk, G. Hammouri, and B. Sunar, "Physical Unclonable Function with Tristate Buffers," in *Proc. ISCAS*, pp. 3194–3197, 2008.
13. A. Maiti and P. Schaumont, "Improved Ring Oscillator PUF: An FPGA-Friendly Secure Primitive," *IACR Journal of Cryptology, special issue on Secure Hardware*, 2011.
14. F. Koushanfar "Hardware Metering: A Survey," *http://aceslab.org/sites/default/files/05-fk-metering.pdf*.
15. J. Roy, F. Koushanfar, and I. Markov, "EPIC: Ending Piracy of Integrated Circuits," in proc. *DATE08*, pp. 1069–1074, 2008.
16. A. Baumgarten, A. Tyagi, and J. Zambreno, "Preventing IC Piracy Using Reconfigurable Logic Barriers," *IEEE Design & Test of Computers*, 2010.
17. T. Kim, R. Persaud, and C. H. Kim, "Silicon Odometer: An On-Chip Reliability Monitor for Measuring Frequency Degradation of Digital Circuits," *IEEE Journal of Solid-State Circuits*, pp. 974–880, 2008.
18. J. Keane, X. Wang, D. Persaud, and C.H. Kim, "An All-In-One Silicon Odometer for Separately Monitoring HCI, BTI, and TDDB," *IEEE Journal of Solid-State Circuits*, pp. 817–829, 2010.
19. Xuehui Zhang and Mohammad Tehranipoor, "Path-delay Fingerprinting for Identification of Recovered ICs," IEEE Int. Symposium on Defect and Fault Tolerance in VLSI and Nanotechnology Systems (DFT), 2012.
20. S. Mahapatra, D. Saha, D. Varghese, and P. B. Kumar, "On the Generation and Recovery of Interface Traps in MOSFETs Subjected to NBTI, FN, and HCI Stress," *IEEE Trans. on Electron Devices,* vol. 53, no. 7, pp. 1583–1592, 2006.
21. K. Uwasawa, T. Yamamoto, and T. Mogami, "A New Degradation Mode of Scaled P+ Polysilicon Gate P-MOSFETs Induced by Bias Temperature Instability," in *Proc. Int. Electron Devices Meeting*, pp. 871–874, 1995.
22. P. Heremans, R. Bellens, G. Groeseneken, and H. E. Maes, "Consistent Model for the Hot Carrier Degradation in N-Channel and P-Channel MOSFETs," *IEEE Trans. Electron Devices*, vol. 35, no. 12, pp. 2194–2209, 1988.
23. Y. Wang, S. Cotofana, and L. Fang "A unified Aging Model of NBTI and HCI Degradation towards Lifetime Reliability Management for Nanoscale MOSFET Circuits," *IEEE Int. Symposium on Nanoscale Architecture,* 2011.
24. *http://www.synopsys.com/Community/UniversityProgram/Pages/Library.aspx*.
25. *http://www.eetimes.com/design/memory-design/4376742/Anti-fuse-memory-provides-robust-secure-NVM-option*.
26. *http://www.sidense.com/technology.html*.
27. *http://www.kilopass.com/products/otp-memory-ip/xpm-otp-nvm/*.
28. N. Reddy, S. Wang, L. Winemberg, and M. Tehranipoor, " Experimental Analysis for Aging in Integrated Circuits," *IEEE North Atlantic Test Workshop (NATW)*, 2011.

Chapter 11
Counterfeit ICs: Path-Delay Fingerprinting

Several light-weight on-chip sensors have been presented to detect recycled ICs based on the degradation of ring oscillators and usage time. They are very effective for recycled ICs detection. However, they only work best for designs with these sensors but cannot address detection of existing and legacy ICs that have no such sensors embedded in them. In order to address this issue, a new technique is proposed based on side-channel information degradation due to aging, including path delay degradation, leakage current degradation, and transient current degradation.

This chapter will focus on the path-delay fingerprinting flow presented in [8]. For new ICs, the delay distribution of paths will be within a certain range. The fingerprint of the new ICs can be generated during manufacturing test of these ICs and stored in a secure memory for future use when identifying recycled ICs. Due to aging effects, such as NBTI and HCI, the path delays in recycled ICs will be larger than those in new ICs. For a chip under authentication (CUA), the larger the path delays are, the higher the probability there is that the CUA has been used and is a recycled IC. A fingerprinting and authentication flow is presented for accurately identifying recycled ICs in [8]. Statistical data analysis is used to distinguish the path delay changes caused by process and temperature variations from those caused by aging. Since the path delay information is measured during the manufacturing test process, no extra hardware circuitry is required for this technique.

11.1 Path-Delay Degradation Analysis

When a chip is used in the field, aging effects could cause some of its parameters to shift over time. NBTI increases the absolute value of the PMOS threshold voltage and results in decreasing transistor current and increasing gate delay [1–3]. HCI creates traps at the silicon substrate/gate dielectric interface, as well as dielectric bulk traps, and therefore degrades device characteristics including voltage threshold

M. Tehranipoor et al., *Integrated Circuit Authentication: Hardware Trojans and Counterfeit Detection*, DOI 10.1007/978-3-319-00816-5_11,

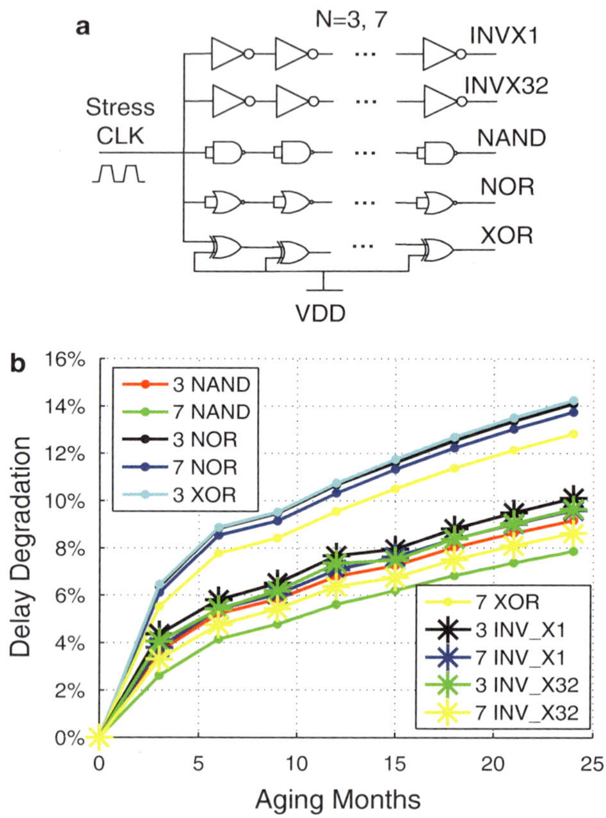

Fig. 11.1 (**a**) An illustrative circuit with NAND, NOR, XOR, and INV chains and (**b**) Delay degradation of the chains

[1–3]. Since recycled ICs have been impacted by all of these aging effects, the path delay of recycled ICs will be different from those of new ICs.

In order to demonstrate the impact of aging on path delay in ICs, in a simple manner, different gate chains were simulated using a 45nm technology [4] as shown in Fig. 11.1a. The simulation was conducted by HSPICE MOSRA [5] with the built-in aging model [5] and combined NBTI and HCI aging effects at a temperature of 25°C. Standard threshold voltage (SVT) INVX1, INVX32, NAND, NOR, and XOR gate chains of different lengths were simulated for up to 2 years of usage. Figure 11.1a shows that all chains are experiencing stress from a 500MHz clock. Any other stress (e.g., DC stress which is a constant "0" or "1", or AC stress with different duty ratios) and usage time could be used in the simulation. Figure 11.1b presents the delay degradation caused by 2 years (24 months) of aging. As can be seen in the figure, different gate chains age at slightly different rates, which depends on the structure of the gates. The XOR gate chain has the fastest aging rate amongst these chains. By comparing the delay degradation rates of the INVX1 and INVX32

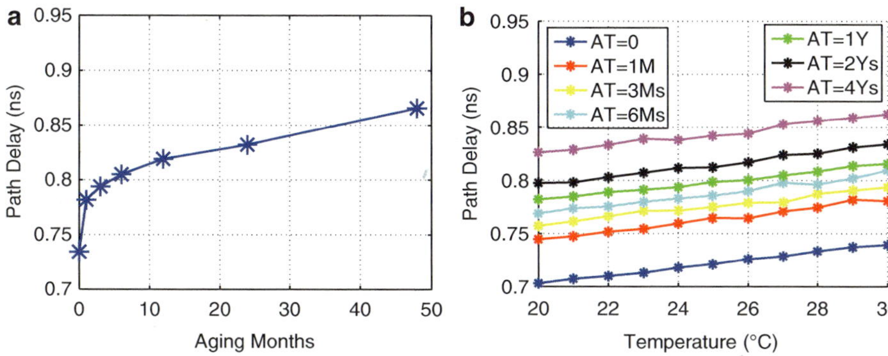

Fig. 11.2 (**a**) Delay degradation of path P_i and (**b**) P_i delay increases with increased temperature

chains, it can be concluded that larger gates will age at a lower rate than smaller gates with the same stress. In addition, the workload (input value and the switching frequency of each gate) also has a significant impact on the aging rate. ICs may be recovered from different used boards from different users who may have applied different workloads to the IC at different times. It is practically impossible to know the exact input vectors applied by the user.

Figure 11.2a shows the delay of a randomly selected critical path P_i (this path includes 22 gates) from the ISCAS'89 benchmark s38417 with stress from a random workload (functional patterns) applied to the primary inputs. The path was aged for 4 years with NBTI and HCI effects at room temperature 25°C. As can be seen in the figure, the degradation of path P_i used for 1 year is around 10 % while if the circuit is used for 4 years, the degradation is about 17 %, indicating that most aging occurred at the early usage phase of the design. Therefore, if there are no environmental or process variations, such degradation should provide great opportunities to identify recycled ICs by measuring one path delay from the circuit. However, these variations have a significant impact on the path delay. On the other hand, different paths age at different rates as demonstrated earlier. Figure 11.2b shows the delay of path P_i under different temperatures at different aging times. In the figure, AT denotes aging time, M represent months, and Y denotes years. As can be seen in Fig. 11.2b, the delay of path P_i increases as the temperature increasing and paths age at different speed under different temperature.

To analyze variations' impact on P_i's delay, Monte Carlo (MC) simulation are run using HSPICE on s38417. 300 MC simulation results of P_i at 25°C are shown in Fig. 11.3a, with 3-sigma 2 % T_{ox}, 5 % V_{th}, and 5 % L inter-die and 1 % T_{ox}, 5 % V_{th}, and 5 % L intra-die process variations. P_i's delay varies around 12 % due to process variations. In addition, process variations also have a significant impact on the aging rate of path delay, as shown in Fig. 11.3b. P_i's delay degradation in the 300 ICs varied around 8 % (4 % $\sim$ 12 %) for 1 year of aging. *These variations evidently make the detection difficult, thus, the path delay shifts caused by aging effects in recycled ICs must be separated from those caused by process variations in new ICs.*

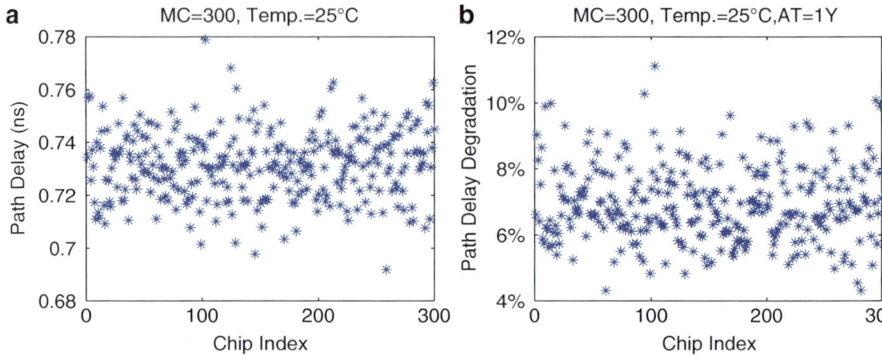

Fig. 11.3 (**a**) Delay of path P_i with process variations and (**b**) Delay degradation of path P_i changing with process variations

11.2 Path-Delay Fingerprinting Considering Aging

Figure 11.4 shows the flow for identifying recycled ICs using path-delay fingerprints and statistical analysis [8]. The proposed flow is divided into three major steps. First, paths are simulated and selected according to their aging rate. Next, the delay information of these paths are measured by a clock sweeping technique in new ICs (either during manufacturing test on all ICs or during authentication on a sample of new ICs) and in any available CUAs. Finally, statistical analysis is used to decide whether the CUAs are recycled ICs or not.

- **Step 1. Path Selection:** Due to the large number of critical and long paths in a circuit, in this step, paths which age at faster rates are selected by analyzing the gate types in different paths and simulating the circuit with different workloads. Paths with higher rates of aging are preferred for fingerprint generation, since the differences in the delay of those paths between recycled ICs and new ICs will be much larger than the differences in paths which degrade slower. Fingerprints generated by fast-aging paths could help identify recycled ICs used for a shorter time. However, there are several parameters impacting the aging rate of a path, including the type of gates composing the path and the workload. Based on these parameters, and the observations made from simulation shown in Fig. 11.1, the following rules are used to select fast-aging paths: (i) paths with more fast-aging gates, such as NOR or XOR gates, will be selected, and (ii) paths that experience more zeros and more switching activity will be selected. More zeros in the path will increase the effect of NBTI on the PMOS transistors, and a high switching frequency will increase the HCI effects on gates, increasing the path delay degradation more significantly.

  Paths with more fast-aging gates would be identified by analyzing the type of gates composing the paths. However, it is very difficult to identify paths that experience more zeros and more switching activity without knowing the specific

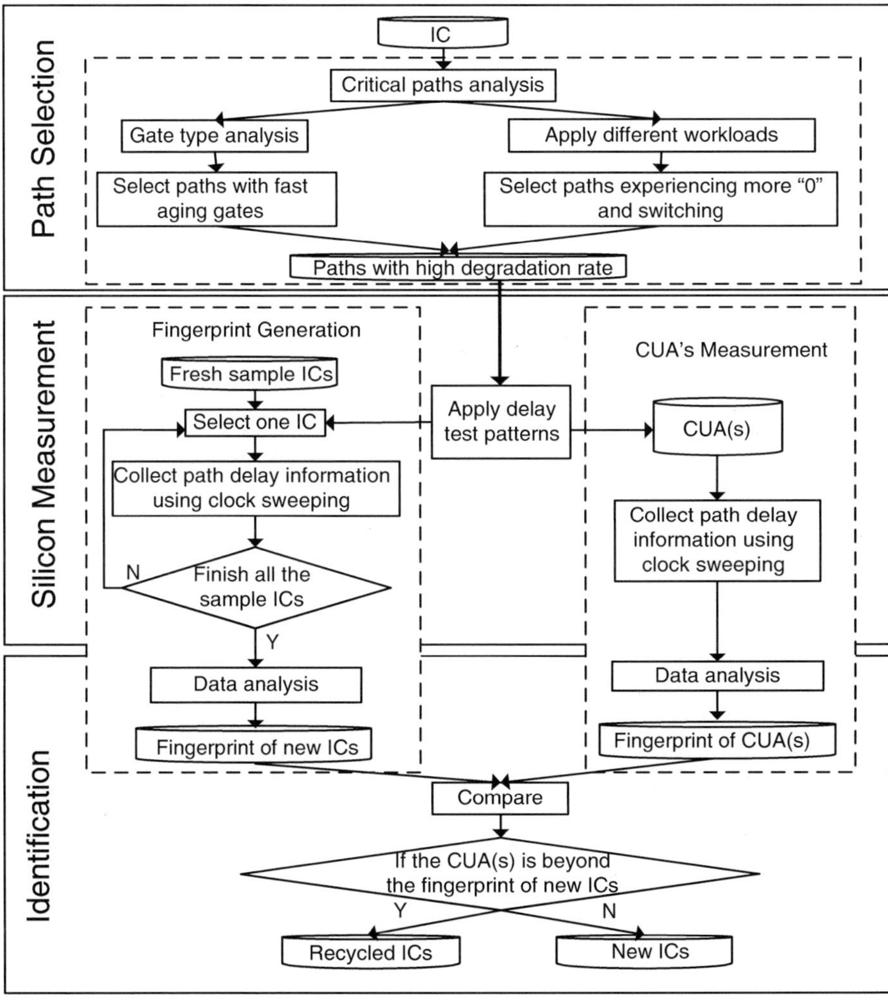

Fig. 11.4 Recycled IC identification flow

workload. Therefore, in the work, different workloads (input combinations) are applied to ICs' primary inputs during logic simulation. For each gate on a critical path, the average switching activity and the zeros it has experienced are calculated. Paths with more switching activity and zeros are then selected using the flow. These paths, along with those composed of the more fast-aging gates, are used to generate fingerprints to identify recycled ICs. The number of selected paths could be adjusted according to the design and its testing procedure. In the simulation, the top 50 paths are selected with fast-aging gates and the top 50 paths experiencing more switching activity and zeros in the benchmark circuit.

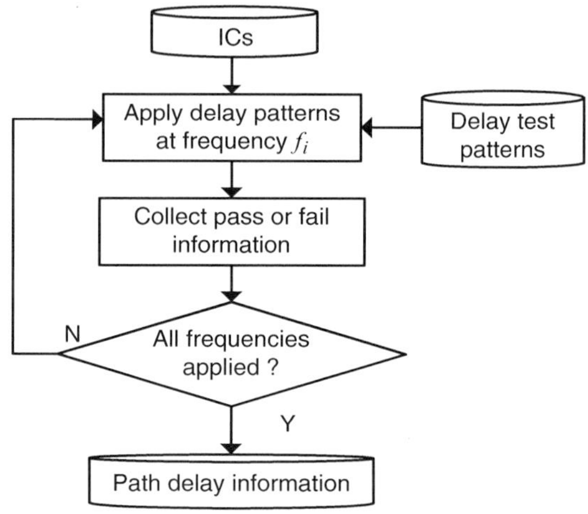

Fig. 11.5 Clock sweeping flow

- **Step 2. Silicon Measurement:** The second step in Fig. 11.4 is to collect the selected paths' delay from the ICs. Note that the fingerprint generation can be done during manufacturing test of a large sample of ICs before shipping them to the market or on a number of new ICs from each production kept by the design house for the purpose of authentication or recycled ICs identification. The larger the size of sample is, the wider of a range of process variations will be included in the fingerprint, reducing the probability that new ICs with large process variations are identified as recycled ICs. Path delay information from the new ICs is measured by performing test procedures on the ICs. Traditionally, test patterns are generated by ATPG before fabrication to detect path and transition delay faults. These patterns will be applied to all new ICs using clock sweeping techniques [6] to measure the path delay of the targeted paths. Note that using clock sweeping is a common practice in industry for speed binning of ICs [6].

  Figure 11.5 shows the flow of the clock sweeping technique. The delay test patterns are applied to ICs at different clock frequencies ($f_1, f_2, \ldots f_n$). Under different frequencies, the paths could pass or fail. If the time period t_i of the frequency f_i ($t_i = \frac{1}{f_i}$) is larger than the path delay, the path will pass. Otherwise, the path will fail. When a path fails, the largest passing frequency will determine the path delay. The frequency step size ($\Delta f = f_i - f_{i-1}$), which depends on the tester, will determine the accuracy of path delay measurement results of silicon chips. For example, with the Ocelot ZFP tester [7], the main frequency is 400 MHz and the frequency step size is 1 MHz. In the simulation, a 5 MHz step size around 1.0 GHz circuit frequency is used for the clock sweeping technique. The measurement environment should keep the temperature as stable as possible, which can be controlled by the manufacturing test environment.

- **Step 3. Identification:** Once the path delay in all sample chips are measured, statistical data analysis will be used to generate a fingerprint for new ICs. For a CUA taken from the market, the same test patterns will be applied in a near-identical environment. The path delay information of the CUA will be processed by the same statistical data analysis methods. In a simple analysis, if the fingerprint of the CUA is outside of the range of the new ICs' fingerprint, there is a high probability that the CUA is a recycled IC. Otherwise, the CUA is likely a new IC. The longer the CUA has been used, the more aging effects it will have experienced, making it easier to identify.

  Without extra hardware circuitry embedded into the ICs, the recycled IC identification technique imposes no area or power overhead. It provides a negligible test time overhead during manufacturing test on a sample of ICs, since only a few patterns must be applied several times at different frequencies. Also, there is no change in the current IC design and test flow since there is no additional circuitry in the IC used for detection. In addition, this method is resilient to tampering attacks. It is inherently difficult for recyclers to mask the impact of aging on the recycled ICs' path-delay fingerprints during the recycling process.

11.3 Statistical Data Analysis

Two statistical data analysis methods are used: simple outlier analysis (SOA), and principal component analysis (PCA). When performing SOA, a single path is randomly selected from the selected path set, and use its delay range in new ICs to generate a fingerprint. The process variations of the CUA may or may not be the same as those within the sample ICs. The selected path delay of the CUA and sample ICs will follow the same distribution, which makes SOA effective in certain conditions. However, a single-path based analysis will not be very effective, due to the limited aging information collected. In general, SOA is expected to be effective in distinguishing recycled ICs used for a long time from new ICs with small process variations.

In order to improve the effectiveness of the technique, PCA is also used to separate the aging effects on path delay from process variations. The path delay information of all selected paths, which have been measured by clock sweeping, will be processed by PCA. In the simulations, the top 100 paths with faster aging rates were selected to generate fingerprints. The delay of each path is one of the variables for PCA to use. Therefore, with N ICs, the dimension of the data set for PCA to generate fingerprint is $N \times 100$. The first three components of PCA in all new ICs were plotted, and a convex hull was created as the fingerprint for new chips. The path delay information of the CUA was also analyzed by the same process and plotted in the same figure. If the CUA is outside of the convex created by the new ICs, there is a high probability that the CUA is a recycled IC.

11.4 Results and Analysis

In order to verify the effectiveness of the recycled IC identification flow and data analysis methods, they were implemented using 45nm technology on a few benchmarks. HSPICE MOSRA [5] is used to simulate the effects of aging on the path delay of different benchmarks. The supply voltage of the 45nm technology is 1.1V. Random workloads (random functional input patterns) were applied to several ISCAS'89 benchmarks. Path delay information was collected using clock sweeping at different aging times. Different process and temperature variations were also simulated to analyze their impact on the effectiveness of the recycled IC identification method.

11.4.1 Process and Temperature Variations Analysis

Table 11.1 shows the three process variations rates used in the simulations. Moving from PV0 to PV2, inter-die and intra-die variations both become larger. PV1 represents a realistic rate of process variations that a foundry might have. Four sets of MC simulation (MCS) were run using different levels of variations, as shown in Table 11.2. For each set of MCS, 300 MC simulations were run to generate 300 chips. During the simulations, the aging effects of NBTI and HCI were simulated with random stress for the benchmark s38417. From the top 500 paths, the paths P_1, $P_2,\ldots,$ P_{50} with fast-aging gates and the paths $P_{51}, P_{52}, \ldots, P_{100}$ with more zeros and higher switching activities were selected to generate fingerprints.

Analysis Using SOA: First, 300 MC simulations were run in MCS1. The maximum aging time is 2 years. Here, SOA was used to process the path delay information. 3 paths (P_1, P_2, and P_{51}) were selected to show the results of SOA. Figure 11.6a–c show the path delay distribution of the 3 paths from 300 ICs used for different aging times. Similar results were obtained for the other 97 paths as well. For each path, the range of the path delay at AT="0" is the fingerprint of the new ICs. If the path delay of the CUA is out of that range, there is a high probability that IC is a recycled one. Note the 300 different MC simulations are used for recycled ICs from those used as sample new ICs. As can be seen in the figure, the delay

Table 11.1 Process variation rates

	Inter-die (3σ)			Intra-die (3σ)		
	V_{th} (%)	L (%)	T_{ox} (%)	V_{th} (%)	L (%)	T_{ox} (%)
PV0	3	3	2	2	2	1
PV1	5	5	2	5	5	1
PV2	8	8	2	7	7	2

Table 11.2 Simulation setup

Experiments	Process variations	Temperature
MCS1	PV0	25°C
MCS2	PV1	25°C
MCS3	PV2	25°C
MCS4	PV1	25°C ±10°C

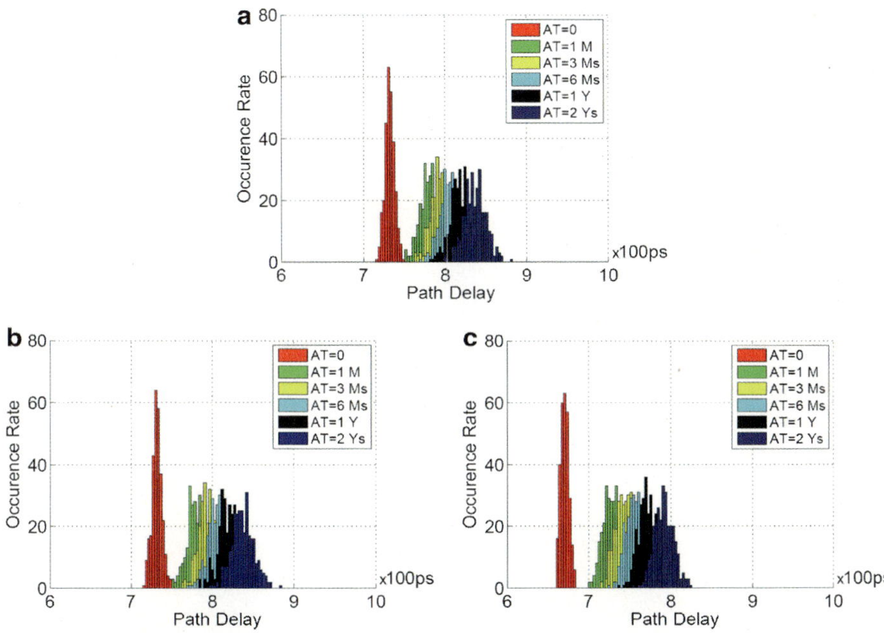

Fig. 11.6 Path delay distribution in ICs with PV0 in MCS1 at different aging times (**a**) Path P_1, (**b**) Path P_2, and (**c**) Path P_{51}

distribution of each path in recycled ICs shifts to the right, relative to the distribution of delays in new ICs. This is because path delay in recycled ICs increases due to aging. The longer the ICs have been used, the more path delay degradation they will have experienced. In addition, the path delay variation increases as the aging time increases. The reason for this is that ICs with different process variations age at different speeds, and the path delay variations become larger as the aging time increases.

Figure 11.6a shows the distribution of path P_1's delay, and the smallest delay of P_1 in recycled ICs used for 1 month is smaller than the largest delay in new ICs. Therefore, the detection rate of recycled ICs used for 1 month is less than 100 % (98.3 %) when using the fingerprint generated by SOA from path P_1. However, the detection rate of recycled ICs used for 3 months or longer is 100 %, which demonstrates that it is easier to detect recycled ICs that have been used for longer amounts of time. If path P_2 is chosen to detect recycled ICs, the detection rate of

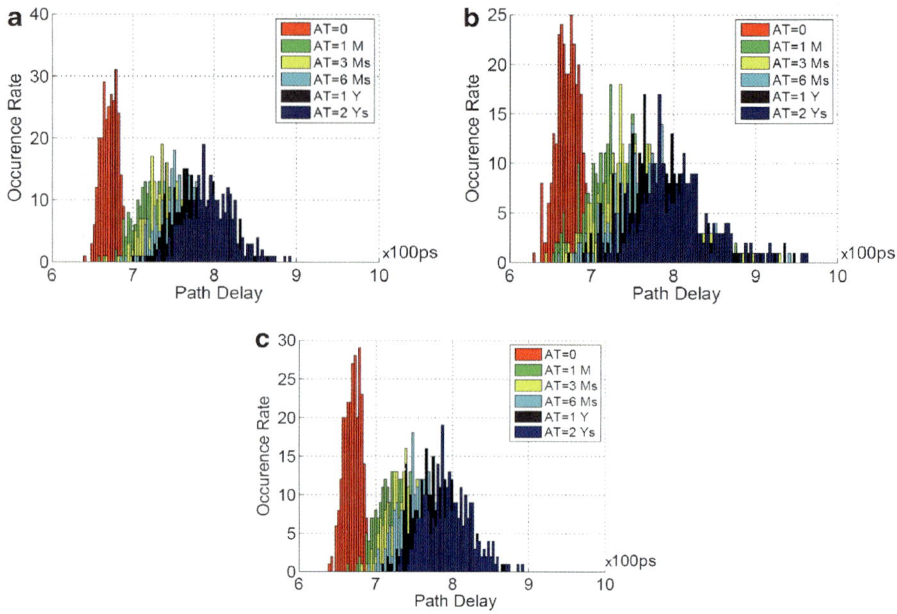

Fig. 11.7 Path P_{51} delay distribution in ICs at different aging times (**a**) in MCS2, (**b**) in MCS3, and (**c**) in MCS4

ICs used for 1 month (95.7 %) is slightly less than when using path P_1. However, if path P_{51} is used, which has the fastest aging rate among the 100 paths, the detection rate is 100 % even if the ICs are only used for 1 month. P_{51} is the most effective path for identifying recycled ICs in this benchmark. From the above analysis, it can be concluded that different paths generate different fingerprints due to their different aging speeds, which makes SOA slightly less effective.

Figure 11.7a and b show the delay distribution of path P_{51} across 300 MC simulations in MCS2 and MCS3. Overall, Figs. 11.6c, 11.7a, and b present the delay distribution of the same path (P_{51}) in ICs with different process variations. By comparing these figures, it can be concluded that the larger the process variations are, the larger the path delay variations in new ICs will be, which makes it more difficult to detect recycled ICs. Even when using the most effective path P_{51}, the detection rates of ICs used for 1 month with PV1 and PV2 drop from 100 % with PV0 to 78.0 % and 50.7 %, respectively. A 100 % detection rate could be achieved if the ICs were used for 1 year or longer with PV1, or longer than 2 years with PV2.

300 MC simulations were also run with $\pm 10\,°C$ temperature variation during the aging process in MCS4 as shown in Fig. 11.7c. The measurement temperature is 25 °C. It shows the delay distribution of path P_{51} and the detection rate of ICs used for 1 month using it is 67.7 %. By comparing Fig. 11.7c and a, it can be concluded that the larger the temperature variation is, the larger the path delay variation is, which makes it more difficult to detect recycled ICs.

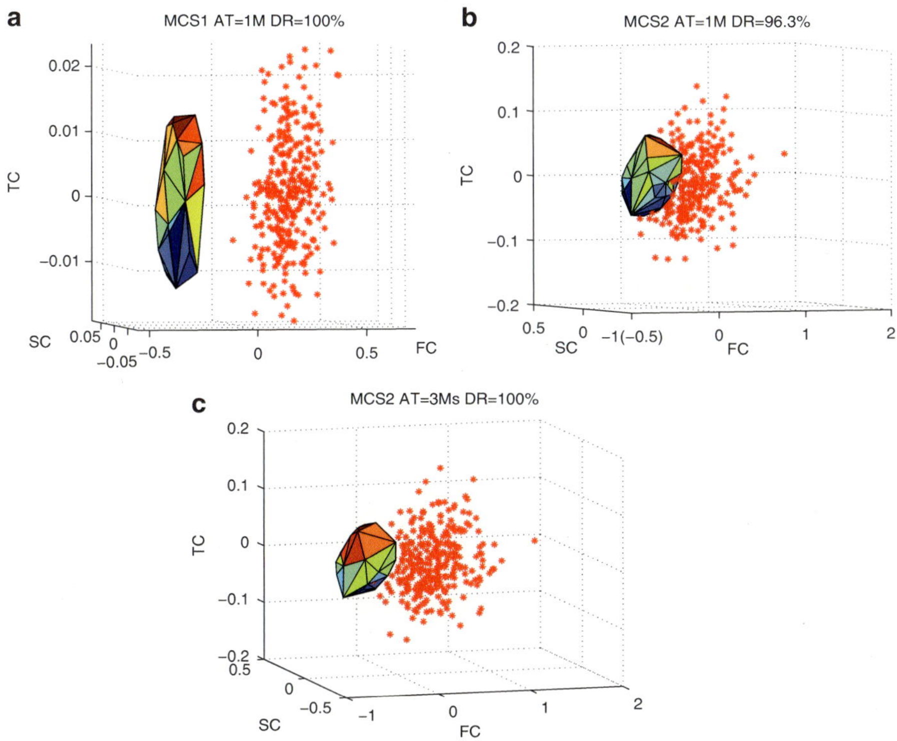

Fig. 11.8 PCA results of ICs under 25 °C (**a**) used for 1 month with PV0 in MCS1, (**b**) used for 1 month with PV1 in MCS2, and (**c**) used for 3 months with PV1 in MCS2

Analysis Using PCA: A similar analysis is done using PCA for different MCSs. Figure 11.8a shows the PCA results of the 100 paths in s38417 with 300 chips in MCS1. FC denotes the first component from PCA, SC represents the second component, TC is the third component, and DR denotes the detection rate. The convex is built up from new IC data, and represents the fingerprint for new ICs. The red asterisks represent chips used for 1 month. As can be seen in the figure, the 300 used ICs were completely separated from the signature of the new ICs. Thus, the detection rate using path delay fingerprints generated by PCA is 100 % for recycled ICs used for 1 month. For recycled ICs used for a longer time, the detection rate will obviously be 100 % as well.

The path delay information from the remaining three sets of MCSs were also analyzed by PCA. Figure 11.8b shows the analysis results of new chips and recycled ICs used for 1 month in MCS2. As can be seen in the figure, some of the recycled ICs are close to the new ICs' fingerprint. The detection rate is 96.3 %, which is much higher than using SOA. By comparing Fig. 11.8b and a, conclusions can be made that (i) the convex hull built up from new ICs in MCS2 is much larger than that in MCS1 (note that the convex hull in MCS1 looks larger than MCS2 due to its small

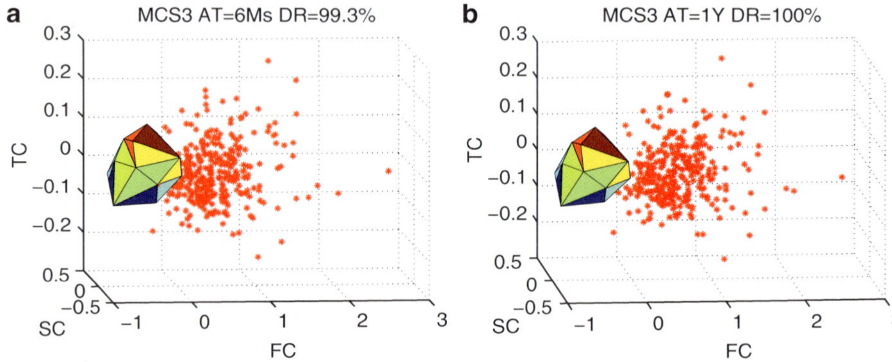

Fig. 11.9 PCA results of ICs with PV2 under 25 °C in MCS3 used for (**a**) 6 months and (**b**) 1 year

scale of axes), and (ii) the recycled ICs in MCS2 are closer to new ICs than those in MCS1, which makes the detection rate in MCS2 less than that in MCS1. The path delay information of 300 ICs used for 3 months in MCS2 were also processed, and the results are shown in Fig. 11.8c. By comparing Fig. 11.8b and c, a conclusion can be made that the longer the chips have been used, the farther they will be from the new ICs' fingerprint. The detection rate of recycled ICs used for 3 months or longer with PV1 at 25 °C is 100 %.

Figure 11.9 shows the PCA results of ICs in MCS3. The detection rate of recycled ICs used for 1 month, 3 months, 6 months, and 1 year are 72.7 %, 89.3 %, 99.3 %, and 100 %, respectively. The figures of PCA results of recycled ICs used for 1 month and 3 months are not shown here since the detection rates are so far from 100 %. Figure 11.9a and b show the new ICs' fingerprint and the recycled ICs used for 6 months and 1 year, respectively. The recycled ICs used for longer times are easier to detect, as seen by comparing Fig. 11.9a and b. By comparing the detection rates in these simulations, it can be concluded that it is more difficult to detect recycled ICs which have higher levels of process variations. The 99.3 % detection rate of ICs used for 6 months and the 100 % detection rate of ICs used for 1 year in MCS3 shows the effectiveness of the technique.

With the same measurement temperature 25 °C, ±10 °C temperature variation is used in MCS4 during the aging process. The detection rate of ICs used for 1 month, 3 months, and 6 months in MCS4 are 90.6 %, 100 %, and 100 %, respectively. The new ICs' fingerprint and the detected recycled ICs used for 3 months and 6 months are shown in Fig. 11.10. By comparing Fig. 11.10a with Fig. 11.8c, a conclusion can be made that the recycled ICs used for 3 months in MCS4 are closer to the fingerprint than recycled ICs used for 3 months in MCS2. This phenomenon demonstrates that temperature variations could increase the path delay variations in new ICs and make it more difficult to detect recycled ICs. However, the 100 % detection rates of ICs used for 6 months in MCS4 demonstrates the effectiveness of the method with process and temperature variations.

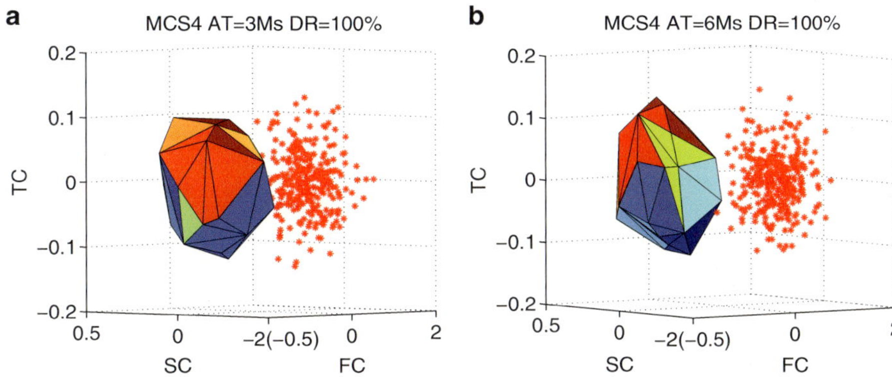

Fig. 11.10 PCA results of ICs with PV1 and ±10 °C temperature variations in MCS4 used (**a**) 3 months, and (**b**) 6 months

Table 11.3 Recycled IC detection rates for s38417

	SOA				PCA			
	1M (%)	3M (%)	6M (%)	1Y (%)	1M (%)	3M (%)	6M (%)	1Y (%)
MCS1	100	100	100	100	100	100	100	100
MCS2	78	96.7	99.7	100	96.3	100	100	100
MCS3	50.7	76.3	85.3	95.6	72.7	89.3	99.3	100
MCS4	67.7	93.3	98	100	90.6	100	100	100

Figures 11.7–11.10 presented some detailed results relating to using this technique on s38417 with SOA and PCA. Table 11.3, however, tabulates these results in addition to some other results obtained using both statistical analysis approaches. These results clearly demonstrate that PCA is more effective than SOA when it comes to identifying ICs used for shorter periods of time.

11.4.2 Benchmark Analysis

In addition to s38417, the ISCAS'89 benchmarks s9234 and s13027 were also simulated to demonstrate the efficiency of this technique on different designs. The process variation and temperature variation rates used in MCS4 were applied to these two benchmarks. The aging stress causing NBTI and HCI degradation in these benchmarks comes from random workloads. 300 MCS were run for each benchmark for a maximum 2 years of aging. The path selection method was also applied to these benchmarks, and 100 paths from each benchmark were used to run statistical data analysis using PCA.

Table 11.4 shows the recycled IC detection rate for all three benchmarks under MCS4 for up to a year of aging. The detection rate for ICs used for 3 months in the benchmarks s9234 and s13207 is 100 %, which matches the results obtained from s38417.

Table 11.4 Recycled IC detection rates—benchmark comparison under MCS4 using PCA

Benchmark	1M (%)	3M (%)	6M (%)	1Y (%)
s9234	88	100	100	100
s13207	89.6	100	100	100
s38417	90.6	100	100	100

The results shown clearly demonstrate that the recycled IC detection method using a path delay fingerprint generated by PCA is very effective, even in designs with large process and temperature variations.

11.5 Summary

In this chapter, a path-delay fingerprinting technique is presented for recycled IC identification. The path delay signatures from recycled ICs are different from those from new ICs due to aging. With no additional hardware circuitry required, this method provides no overhead on area and power consumption. The simulation results of different benchmarks with different process and temperature variations demonstrated the effectiveness of the method.

References

1. S. Mahapatra, D. Saha, D. Varghese, and P. B. Kumar, "On the Generation and Recovery of Interface Traps in MOSFETs Subjected to NBTI, FN, and HCI Stress," *IEEE Trans. on Electron Devices,* vol. 53, no. 7, pp. 1583–1592, 2006.
2. K. Uwasawa, T. Yamamoto, and T. Mogami, "A New Degradation Mode of Scaled P+ Polysilicon Gate P-MOSFETs Induced by Bias Temperature Instability," in *Proc. Int. Electron Devices Meeting*, pp. 871–874, 1995.
3. P. Heremans, R. Bellens, G. Groeseneken, and H. E. Maes, "Consistent Model for the Hot Carrier Degradation in N-Channel and P-Channel MOSFETs," *IEEE Trans. Electron Devices*, vol. 35, no. 12, pp. 2194–2209, 1988.
4. *http://www.nangate.com/?page_id=22.*
5. *Synopsys, HSPICE user guide, 2010.*
6. J. Lee, I. Park, and J. McCluskey "Error Sequency Analysis," *Proc. VLSI Test Symp.*, 2008.
7. INOVYS, *http://www.etesters.com/listing/40e8f648-a2d6-23b8-949b-4b3c005c86fb/Ocelot_ZFP_-_Test_System_for_Complex_SOCs.*
8. Xuehui Zhang, Kan Xiao, and Mohammad Tehranipoor, "Identification of Recovered ICs using Fingerprints from a Light-Weight on-chip Sensor", Design Automation Conference (DAC), 2012.

Index

M. Tehranipoor et al., *Integrated Circuit Authentication: Hardware Trojans and Counterfeit Detection*, DOI 10.1007/978-3-319-00816-5,